A JOURNEY

THROUGH THE ELEMENTS

A JOURNEY THROUGH THE ELEMENTS

Memoirs of a Fortunate Geologist

Garry Lowder

Connor Court Publishing

Published in 2019 by Connor Court Publishing Pty Ltd

Copyright © Garry Lowder 2019

All rights reserved. No part of this book may be reproduced or transmitted in any form or by any means, electronic or mechanical, including photocopying, recording or by any information storage and retrieval system, without prior permission in writing from the publisher.

Connor Court Publishing Pty Ltd
PO Box 7257
Redland Bay QLD 4165

sales@connorcourt.com
www.connorcourtpublishing.com.au
Phone 0497 900 685

ISBN: 978-1-925826-65-4

Front Cover Design: Maria Giordano

Printed in Australia

CONTENTS

Foreword 7
1. An Elemental Beginning 9

1. FIAT LUX

2. Berkeley in the Sixties 19
3. Thank You, Mr Ford 31
4. Golden Gate 41
5. Black Glass 49
6. Dry Fly 63
7. Wriggling Rainbows 73
8. Reflections 85

2. "MASTA BILONG BRUKIM STON"

9. Rabaul 91
10. The Talasea Club 103
11. Volupai 113
12. Bulu Muri 123
13. Lake Dakataua 131
14. Makalia Conquered 143
15. The Prof 157
16. The Masalai 169
17. Bulu Dava 179
18. Sleeping Giants 185
19. Taim Bilong Go Pinis 199

3. EQUATORIAL EMERALDS

   20. Six Thousand Islands — 205

   21. Getting There is Half the Fun — 207

   22. Sungai Ular Besar — 219

   23. Kayubulan Ridge — 231

4. The PASSIONATE PROSPECTOR

   24. No Stone Unturned — 245

   25. Science *vs* Serendipity — 251

   26. Discovery — 257

   27. The Golden Mile — 263

   28. Naked Heikki — 275

   29. Rainforest Realtor — 283

   30. The Dark Side — 289

5. CONTINUING PASSIONS

   31. A Life Cycles — 305

   32. Rainforests — 307

   33. Mountains of Fire — 315

   34. Mutnovsky — 321

   35. Valley of the Geysers — 327

   36. Hawaii — 335

   37. Outback Elegy — 341

   38. Journey's End — 357

      Acknowledgements — 363

FOREWORD

A Journey through the Elements is an excellent read for those in all walks of life. It is more than the memories of a brilliant geologist, in that his descriptive style of writing transports you into Garry Lowder's life and his progression from University student, to junior geologist, to senior explorer and finally to general manager and company chairman. Along the way you feel the excitement and trepidation involved in undertaking graduate research as a PhD student at one of the top Earth Science schools in America, and learn of the strong relationship developed between professor and student, not only in the laboratory, but also in the field. Garry and Margaret were a young and newly married couple when Garry undertook his field studies in the jungles of Papua New Guinea, and their stories of living with the local natives, struggling with transportation and climbing active volcanoes are the stuff of the adventures of Indiana Jones.

Not only is this an entertaining and amusing personal memoir, it is also highly educational, delivering the author's insightful view of geological processes that have shaped the Earth through time and which continue to this day. Erupting volcanoes, earthquakes, magmatic intrusions and the processes involved in forming rich deposits of gold and copper are weaved into the narrative in such a way that they become an adventure.

The excitement of the quest for minerals in the rain forests of Papua New Guinea and Indonesia, the pastoral country of eastern Australia and outback bull dust country of Western Australia are vividly expressed in the pages of Garry's life, and give the reader a sense of being part of the hunt. However, this book is not all about adventure and success, but gives a warts and all account of the fulfillment of a geological career, and the difficulties, dangers and setbacks faced by Garry along the way.

A Journey through the Elements presents an "insider's" view of the world around us that will resonate strongly with anyone interested in natural history. As readers are transported into tropical rainforests and arid deserts, or across mountain ranges and sweeping plains, they encounter the Elements – Earth, Air, Fire and Water – in a way that brings the natural world to life in an intense and personal way.

Whether you are a student of the natural sciences, practicing geologist, retired professional or simply someone happy to read about an exciting life spent actively engaging with the natural environment, I can strongly recommend this memoir for your enjoyment.

Professor Ross Large AO

Professor Emeritus

University of Tasmania

1

AN ELEMENTAL BEGINNING

> "It's not what you look at that matters, it's what you see."
>
> Henry David Thoreau
> American writer and polymath
> 1817 - 1862

When my mother told her life-long friend, Esma, that her son was studying geology, Esma replied:

"Oh, you mean he's going into the church?"

Esma's confusion of geology with theology knocked my youthful ego back a peg or two, deservedly I admit. It is amusing in retrospect but is perhaps understandable for a woman born one hundred years ago, who had only a limited education as she grew up during the Great Depression. In the early 1960s, geology as a science was virtually unknown outside academic circles. Professional geologists were rare specimens indeed. Surely no one today would make Esma's mistake. Perhaps not, but to my mind, one of the great failings of modern society is its ignorance of the earth we live on. There are many people who would "save the planet" but few who understand just how that planet really works. Nor do they appreciate the length and resilience of its vast history and why, therefore, it does not really need us to save it. To a large extent, this is because the would-be planet-rescuers' view of the world is limited to two dimensions, when even three dimensions are not enough. They look around themselves but not below themselves; they look at the world but do not see it. It is only when we read the stories of creation that are written in the rocks beneath our feet that our place on earth can be seen in context. While most people look at a rock and see ... well ... a rock, I look at a rock and see a story. By reading the tales of trauma and triumph that are emblazoned on every outcrop we begin to appreciate the vastness of Einstein's fourth dimension – time – and why we need that extra dimension to really understand our planet.

It was from that background that I ventured into earth science at Sydney University in 1961 at the tender age of 16 (and a half). It was the beginning of a life-long fascination with our planet and a long life immeasurably enriched by the people, places and experiences that more than fifty years as a geologist has exposed me to. Over those years my zeal for the earth has taken me to countless distant and exotic parts of the world, both in my native land of Australia and in foreign realms. In those places, I have met some wonderful people and had numerous adventures and surprising encounters – some hazardous and frightening, some joyous and rewarding. Together these people and the experiences I shared with them have shaped me as a person and formed an awareness of environment that makes me extremely grateful that it was in fact geology, and not theology, that became my life's work.

I have read the stones and listened to their voices murmuring in the desert, in the rain forest, in the mountains and on the plains. It was the stones that told me how the world was made; it was the stones that explained why the landscape looks like it does; and it was the stones that showed me where the earth's treasures are hidden. I have found a few of those treasures over the years and watched respectfully and gratefully as some of them were extracted. Slowly, I have come to understand the importance of time in forming a holistic view of the world. My communion with the rocks has revealed the unhurried pace of the earth and the immense history that those stones, those rocky outcrops, those pebbles beneath the gurgling waters, hold in their embrace. They are the papyruses on which the story of Planet Earth was inscribed eons ago and they now lie waiting for us to discover them, if only we will take the time to look and listen.

A life so enriched should be shared and it has indeed been shared with a select group, not least my wife, Margaret, who has been with me in some of my adventures, and the many colleagues I have worked with. For my children, it has been more removed; mythical tales of their father's adventures during his many absences from their formative years. As I look back over my life I can easily see how mythology arises, for many of the encounters I have had could not be repeated today and to some people they are scarcely believable. Yet they are real and I am driven to record them now, both for myself and for the sake of those whose lives may have been more conventional, but who wish they hadn't been. It is my fervent hope that in reading this memoir my audience will share in my good fortune, be entertained by my story-telling and come to share my passion for the earth we live on.

The story begins (**Fiat Lux**) at the University of California, Berkeley, where I completed a PhD in 1970. From the glaciated Sierra Nevada mountains to the lush high plateaus of Utah, from the depths of the Grand Canyon to the peaks of the Canadian Rockies, the sheer extravagance of Western American geology etched an infatuation with the natural world deeply into my soul. It was at Berkeley that I met Lindley Sale, the personification of American generosity and one of life's true gentlemen. And it was at Berkeley that Ian Carmichael, the personification of academic excellence, came into my life. Ian was my PhD supervisor and was very much a larger-than-life character who inspired me to achieve way beyond my own expectations. In doing so he became the most influential person in my life.

Then the scene changes to Papua New Guinea (**Masta Bilong Brukim Ston**), where I took a break from Berkeley to conduct field work for my PhD, studying geologically young volcanoes, ably assisted by my new wife, Margaret, a beautiful woman of high intelligence, strong character and deep loyalty. The experiences Margaret and I shared by working in a remote part of this remarkable but today much troubled country profoundly affected both of us. There we were immersed into the vibrancy of the rain forest, the geological dynamism of the Pacific Margin and the joyous humanity of the Melanesian people, whose lives we shared by residing in their villages for weeks at a time.

Fortified by the discipline instilled into me at Berkeley and supported by Margaret, I came back to Australia in 1970 and launched into the mainstream of life as a geologist. Before very long I found myself working in the Indonesian rain forests, where my love of nature was indulged most lavishly and I was endowed with a deep respect for this archipelago of **Equatorial Emeralds**. There it was that I had a startling confrontation with hostile monkeys while lost in the jungle. There too I survived being adrift in a dugout canoe while an off-shore breeze pushed us further and further out into the blackness of a Moluccan Sea night.

Then come stories from subsequent years that show why I regard myself as a **Passionate Prospector**. The scenes range from sweating in the PNG rain forest to sweating in a Finnish sauna, from exploring dark tunnels deep beneath the Golden Mile to enduring my own *"Yes Minister"* experience. With my feet back on the ground I also provide some clues on how to find a gold mine and an account of when I did just that.

Finally, I share my **Continuing Passions,** revealing how geology, love of nature and appreciation of my fellow man have made me who I am. My life as a geologist has shown me that patiently, at its own pace, the earth embraces change and endures cycles just as we do, but it does so in a timeframe that is humbling when compared with our own fleeting existence. To understand this is to know our place in the cosmos.

***

My geological journey began more than five decades ago, which is quite a long time in human terms, though it is but a flicker in the current of life on earth, which started flowing over three thousand five hundred million years ago. If all of earth history were equated to one 24-hour day, my lifetime would be less than one thousandth of a second. In that same equation, the first life on earth would appear at 6.10 am, the first animals with hard parts (shells) would not show up till 9.05 pm and the dinosaurs would finally meet their doom as late as 11.40 pm. Homo sapiens, us, would appear just three seconds before midnight. Our species is thus very much the "Johnny-come-lately" of Planet Earth. And yet we think of it as ours to exploit, ours to destroy, or ours to save!

In high school, I had no exposure to geology but was good at science, which then meant mathematics, physics and chemistry. As my final year of high school progressed I faced a decision about my tertiary education. No one in my extended family had till then attended university but my hard-working parents believed strongly that education was paramount. They encouraged me to seek a university education when it became obvious that I had the academic aptitude to do so. But what to study?

"I think you would make a good science teacher," my father

suggested, recognising my talent for the subject but limited in his knowledge of scientific careers.

My mother was less committed:

"Well I've always wanted one of my boys to be a policeman," was her offering.

I was struggling to see myself in a white coat, surrounded by laboratory paraphernalia and squabbling children. I found it even more difficult to imagine myself in a blue uniform. There had to be another option.

I had been a boy scout and loved the outdoors so it was not surprising that my attention eventually turned to the science of geology, then pretty much a clean slate to me, as it was to most people. My older brother, Graham, had begun his long career in commercial aviation by piloting small aircraft around the mountains and through the mists of Papua New Guinea. His depictions of the volcanoes he saw as he flew by in his little Cessna had intrigued me deeply. This created an unsated appetite that I thought might be satisfied, together with the need to support myself, by becoming a geologist, combining science, the great outdoors and expenses-paid travel to far-away places.

The more I considered it, the more my curiosity for the earth grew. I wanted not just to see the sights of the outback and foreign lands but also to understand them, to know how they got to be as they are. I wanted to comprehend and experience at first hand the fundamental nature and history of the earth. From the beginning, I wanted to know what stuff the earth is made of.

So did the Greeks and other ancient societies. As early as 450BCE Greek philosophers had formulated an hypothesis that would hold sway for two thousand years. They deduced that the material form of the world was made up by combinations of four essential substances they called **Elements**. These four elements were identified as **Earth**, **Air**, **Fire** and **Water**.

All matter was thought to comprise mixtures of these four Elements, a view later supported by Aristotle, who added what he regarded as a fifth element that he called **Aether**, which came to be thought of as a kind of glue holding the visible elements together.

The principle of there being four fundamental, indivisible components of matter (or five if you include the aether) inhabited all science and philosophy for the next two millennia. The significance of the quartet was extended to medicine when Hippocrates equated the four elements to the four "humours" that he believed could be found in the human body, which were reflected in the four temperaments of personality. For a very long time, the goal of medicine was to keep the four humours in balance in the body (bleeding one out if it was over-represented) and to match humours with temperaments to maintain good mental and physical health.

It was not until modern physics and chemistry began to develop as rigorous sciences, leading to world-changing discoveries, that the ancient Greek view of matter was discredited. It is interesting, though, to note that even now the four Elements can in a way be correlated with modern concepts of the four states of matter, namely solid (= earth), liquid (= water), gas (= air) and plasma (= fire).

When I think about what a life in geology has meant to me, it is quite remarkable how many of my experiences fall into categories that match rather well the ancient Greek concepts. In so many cases the experience brought with it a component of earth or air or fire or water. For me, the Elements have been a recurrent theme: Earth, Air, Fire and Water, bonded loosely by the Aether. Together, the Elements link this narrative and chart my journey through danger, isolation, awe and excitement, as I bring to life a dynamic and inspiring world that many people cherish but few have encountered quite so vividly.

1. FIAT LUX

2

BERKELEY IN THE SIXTIES

Years ago there was an art-house movie by that title and if I have offended anyone by using it here I apologise. I use this title advisedly though and expressively. It probably does not mean a great deal to younger generations, or to most Australians, but to many Americans, especially those of my own vintage, the phrase evokes the entire lexicon of an era like no other. "Flower Power"; "Haight Ashbury"; "People's Park"; "Free Speech" and "Student Demonstrations": these words bring with them a flood of memories and images that tell of a time when America's happy, post-war innocence began to give way to both a more mature self-awareness and the harsh reality of a non-conforming world.

America was in the ascendency in the 1960s, conquering outer space, dominating world business, and driving innovation and scientific discovery. America was unstoppable, the rich, free and democratic nation to which all others might aspire. True, it was mired in an unwinnable war in Vietnam. And the oppression of its black people was deeply entrenched, especially in the South, where Martin Luther King showed the power of non-violent protest. Until he was violently liquidated in April, 1968, just as John Kennedy had been five years before him and Robert Kennedy was two months after him.

Blessed with abundant funding and a steady flow of baby-boomer adolescents as students, America's universities flourished. The East Coast had long boasted of its great and privileged Ivy League institutions – Princeton, Harvard, Yale, Columbia and four others – while the West Coast played catch up. But in the 1960's the western universities' time was at hand, as westward migration drove strong population growth, especially in California. An explosion of scientific and technological breakthroughs combined with the burgeoning population to feed the development of a new breed of university, most of them state-owned and egalitarian in nature. And the greatest of the new breed was the University of California, which had seven campuses at that time; now it is ten: Los Angeles (UCLA), Davis, Irvine, Merced, Riverside, San Diego, San Francisco, Santa Cruz, Santa Barbara and the oldest and most esteemed of them all – Berkeley.

The University of California at Berkeley (known as UCB, or Cal, or U C Berkeley, or just Berkeley, whose motto is: *"Fiat Lux"* or *"Let there be Light"*) sits on the eastern shore of San Francisco Bay, directly opposite San Francisco's celebrated Golden Gate Bridge. There it nestles up to the wooded Berkeley Hills where Australian eucalyptus trees dominate the skyline. The campus pivots on its iconic campanile, known as Sather Tower, which has a lovely carillon and is widely recognised as a symbol of the institution. The Berkeley campus is an attractively landscaped arena in which to undertake advanced learning. Even with its large Student Body (some 25,000 undergraduates and 10,000 graduate students) it did not feel crowded in my day and the Administration seemed to work efficiently. The main, formal entrance to the campus is on the western side, at the top of University Avenue, which extends from the Eastshore Freeway up to the gate of the University.

The east side of the campus stretches up the hill slopes to the football stadium, botanic garden, the Lawrence Hall of Science and the Lawrence Berkeley Laboratory. It was in this laboratory that, during World War II, Ernest Lawrence, J. Robert Oppenheimer and others established what came to be known as "The Manhattan Project" – the development of the Atomic Bombs that were used at Hiroshima and Nagasaki.

On the south side of the campus Telegraph Avenue runs away from Sather Gate towards Oakland. In the 1960's the first few blocks of this street comprised a motley collection of book shops, record shops, second hand clothing stores, greasy diners and takeaways. It was also the focal point for Berkeley's Hippie culture, which developed in parallel with that in San Francisco, where the intersection of Haight Street with Ashbury Street (the "Haight – Ashbury District") was the locus of action. In 1969 Telegraph Avenue would gain great notoriety when demonstrations against the use by the university of a vacant block it owned as a car park, instead of a "People's Park", led to the mobilisation of the National Guard. As trucks carrying troops rolled down Shattuck Avenue (the main street of Berkeley) the campus was sprayed with tear gas from a helicopter. There was no escape from the gas for those of us on campus that day but the worst outcome of the National Guard involvement was the tragic death of a student by shooting. At that time, Ronald Reagan was Governor of California. Arch conservative that he was, Reagan frequently criticised the culture of Berkeley as a rabble of radicals and "commies" so he was untroubled by the precedent he set in using the National Guard to quell a student protest. Mind you, whether he liked it or not, as Governor, his signature appeared on Berkeley degrees, including my PhD certificate.

Student protests and "commies" aside, Berkeley was a stimulating and rewarding place to live, with a wide range of habitats and cultural styles. At the lower elevations, near the Bayshore freeway, there were the crowded tenement blocks that were mostly occupied by Blacks and Hispanics. Moving upslope, closer to the campus, student housing dominated; sometimes seedy but mostly functional apartment blocks where groups of students from all over the state and elsewhere crowded into the available space. Higher still, with views over San Francisco Bay, were the flash apartments and large houses of the university Faculty[1] and the wealthy, some of whom had factories down amongst the tenements. Overall it was an eclectic mix of wealth and poverty, of privilege and deprivation, of race and religion, of locals and foreigners, of high intellect and the great unwashed. As the local radio station, KPAT, reminded us every day:

"There are more Nobel Prize winners in Berkeley than any other place on earth."

I had arrived in Berkeley in September, 1966, to take up a one-year study Fellowship that had been awarded to me by the Rotary Foundation. The Fellowship paid a generous stipend, including a travel allowance, and was sufficient to cover fees, accommodation and general living costs. In return, I was expected to make myself available as a speaker at Rotary Club meetings in the region from time to time, something I did willingly and soon learned to enjoy. My presentation was a simple story of Australia and its mineral wealth (yes, even in the 1960's there was a mineral boom, culminating in the great Poseidon nickel boom) and this seemed to go down well with my audiences, the largest of which numbered over 2,000 (at a Rotary District Conference). In the 1960's Australia was relatively unknown to most Americans and there were still people who had never heard of us, or at best confused us with Austria. After addressing a Rotary meeting in Berkeley, an audience member approached me and said:

"You speak beautiful English! Where did you learn to speak like that?"

---

[1] In American universities, the term 'Faculty' refers to the academic staff.

When this happened, I accepted the compliment gracefully, realising that this man had no idea what language was native to Australia. In fact, I quickly became quite adept at making small talk and chatting up my hosts in an obliging manner, while keeping a straight face as they showed their lack of awareness about foreign lands. As the fortunate recipient of their generosity, I did not want to insult my hosts by pointing to their naivety. After all, the "*Raison d'être*" of the Rotary Foundation Fellowship was the promotion of "International Understanding". America then, even more than now, was a very self-sufficient society and people like the men I met at Rotary meetings saw no shame in their ignorance of foreign affairs.

Before going to Berkeley my experience of public speaking was very limited. At first, I was a bit intimidated. But once I realised that the only person present who knew what was supposed to come next was me, and that if I made a mistake, only I would know, my nervousness gradually dissipated. The public speaking practice gained during those days has proven to be an invaluable life skill, one still in use today, as I occasionally address Probus and University of the Third Age ("U3A") groups around Sydney.

There was just one downside of the Rotary Foundation Fellowship – it required celibacy! Taking up the award meant leaving my fiancé, Margaret, at home for ten months. This was a great sacrifice on her part and I shall be forever grateful to her for that.

Soon after my arrival in Berkeley I teamed up with a newly met Canadian sociology PhD student, Maurice Manel, to rent an apartment a couple of blocks south of the campus. It was a functional but rather decrepit old building, covered in wooden shingles and full of students. It later burned to the ground, long after we had vacated it fortunately. The apartment itself was basic, with a tiny kitchen, but it was cheap ($120 per month) and met our needs. We had a semi-circular table in the kitchen, with the flat side pushed up against and just above the window ledge. Cleaning up after a meal was simply a matter of sweeping the crumbs out the window and into the garden below. Who said academics were not practical people? We even washed the kitchen floor occasionally too, once the stickiness of the linoleum constituted a serious risk of entrapment.

My daily routine was to stroll up Telegraph Avenue, trying not to stare at the unkempt – and often unwashed – Hippies, to Sather Gate. Once on the campus I crossed past the Hearst Library to the north side where the Earth Sciences Building was situated. This relatively new, five-storey building was my home away from home for four years, from 1966 to 1970. From its upper floors, there is a superb view directly across the Bay to the Golden Gate Bridge. Many was the time I stood on the balcony and watched the renowned San Francisco fog roll in under the Golden Gate and gradually make its way across the Bay to envelop Berkeley. Just in front of the building a large California Coast Redwood tree stood proudly, its foliage seeming to shine through the mist as though lit from within. The fog is a summertime phenomenon and commonly leads to summer days being quite cool and making spring and autumn the best seasons in the Bay Area.

The Earth Sciences Building was home to the Department of Geology and Geophysics and the Department of Paleontology. In Australian universities these two would be rolled into one, but here they were separate, which was convenient when it came to PhD Oral Exam times, when the five-member interview panel had to include two professors from other departments. In the entrance lobby the geophysicists had set up a seismograph, with a needle pen tracing a continuous squiggly line of red ink on a paper-wrapped drum as it slowly revolved under it. The infamous San Andreas Fault passes through San Francisco and was the locus of the devastating 1906 earthquake. A splay [2] from that fault, known as the Hayward Fault, passes through the Berkeley campus, right under the football stadium. Slow creep along the Hayward Fault has displaced the stadium's original walls, built in 1923, by about 30cm. Whenever there was an earth tremor, which was often in this seismically active region, we students would rush down to the lobby to watch the seismograph record the shock. The needle would swing wildly back and forth as the earthquake waves rumbled through, greatly increasing its amplitude and scratching the red ink densely across adjacent, pre-tremor lines. Slowly, the needle would settle down and resume its monotonous squiggle as the drum continued its steady rotation. That seismograph was a very visible clue to the dynamic nature of the earth and a constant reminder that the stability we humans crave is only ever temporary and decays over time. Entropy always increases.

The reason I had chosen Berkeley as the institution in which to take up the Fellowship award was simple – almost all of the geology textbooks I had used as an undergraduate in geology at the University of Sydney had been written by Berkeley professors. At first, I was quite in awe of these famous men (yes, they were all men) as I attended their classes and passed them in the corridor. Names like Turner, Verhoogen, Williams, Gilbert, Weiss, Curtis and Meyer that had been no more than ink on a printed page were now flesh and blood in front of me. Berkeley claimed, I think justifiably, to host the best earth science school in the United States, if not the world, and I was now part of it. But, renowned as these men were, it was not one of them that ended up taking the central role in my graduate education. That distinction fell to a young, relatively unknown but up and coming Associate Professor named Ian Carmichael.

---

[2] i.e. a side branch

Petrology is the branch of geology that deals with the origin, evolution, composition, mineralogy, structure and form of rocks (from the Greek: *'petra'* – rock and *'logos'* – study), and igneous petrology (the study of rocks that were once molten) had been the specialisation pursued during my Honours Year (1964) at Sydney University. The Rotary Fellowship thus gave me the opportunity to pursue graduate studies in igneous petrology as a student of the authors of my undergraduate textbooks. Upon arrival in Berkeley I had expected to be taken under the wing of Professor Frank Turner, at that time probably the world's most renowned living petrologist. How naive and presumptuous that was on my part! Turner was kind but firm:

"You should go and see Professor Carmichael, around the corner in Room 419," he suggested, in a manner that would brook no argument.

That turned out to be the most significant redirection of my life. Ian Carmichael was then perhaps Turner's protégé (although Ian would hate me saying so) but he was totally unknown to me. Over the next four years Carmichael became my supervisor, my mentor, my inspiration, my friend, my examiner, at times my nemesis and certainly the most influential person in my life.

Not that working with Carmichael was easy. On the contrary, his strong and colourful personality, often expressed in equally colourful and exuberant language, challenged many people and made them wary of getting close to him. He certainly did not suffer fools gladly and was not at all averse to calling out incompetence wherever he saw it. The fact that he did so face to face and was never reticent to show the courage of his convictions directly was confronting, especially to aspiring students. It caused many to slink away in humiliation. In my time with Ian I copped plenty of this myself and yet, perversely, my admiration for the man only grew. His words were strong, challenging, at times offensive, but they were perceptive and in the main, constructive and character building. For his criticisms were never without foundation and once I and others who worked with him learned to filter out the noise and hear the real message he was delivering we made great strides forward in terms of academic excellence and our development as scientists. He knew what he was doing and once I also knew what he was doing my relationship with Ian grew way beyond that of teacher and student as we gradually became colleagues and friends.

Only after some time did it become apparent that his confrontational style was but one side of this complex and surprisingly generous man, as more and more his other students and I began to learn of his effusive praise of us to third parties, beyond our hearing. Ian was not exactly a role model, for none of us had his larger-than-life personality, but the standards he inculcated into us and his insistence on nothing but the best lifted each of us above the intellectual plateau on which we might otherwise have been stranded. It became abundantly clear why Ian Carmichael had been appointed to the Faculty at Berkeley and indeed, why the Department of Geology and Geophysics at that great institution had the world-beating reputation that it did.

Ian Carmichael died in 2011, still in Berkeley and aged just 81. Reading an obituary published by the University at that time I was gratified to have my view of him endorsed by two graduate students who studied for their PhDs under Ian some years after me:

> *"Carmichael will be remembered as a mentor to two generations of Ph.D. students in geology, a friend and pioneering volcanologist to Mexican geologists, and a gregarious story-teller with a zest for life to his family and friends. He even has a mineral — carmichaelite, an hydroxyl-bearing titanate*

from Arizona – named in his honor.

"None of his students and colleagues escaped being shaped, in some way, by the hurricane force of Ian's personality: his infectious enthusiasm, his imaginative brainstorming, his intellectual generosity, and his impatience," said Rebecca Lange, his close friend, previous Ph.D. student and chair of the Department of Earth and Environmental Sciences at the University of Michigan. "His exuberant pushing and prodding, combined with his belief in us, forced us to stretch ourselves and realize potentials we never knew we had."

"He was a giant in the field of igneous processes and volcanoes," added Paul Renne, director of the Berkeley Geochronology Center and a professor in residence in the Department of Earth and Planetary Science at UC Berkeley. "He was the leader in application of physical chemistry, in particular thermodynamics, to understanding how magmas originate and evolve."

Ian Carmichael truly was a giant of a man and a major influence on my life. When I first met him, he was only 35 years old, still in the early stages of an academic career that began with a PhD at Imperial College in London (for studying *Thingmuli*, a volcano in Iceland). That was followed by a short stint at the University of Chicago. His appointment to the Faculty at Berkeley came just as his reputation in igneous petrology, and especially volcanology, was beginning to rise. He stayed at Berkeley until he retired. Ian never lost his South London accent (which the Americans incorrectly labelled "Cockney") but he adapted to life in America very well, driving a Ford Mustang Convertible with as much flair as any Hollywood celebrity. He was a man of large stature, always a little overweight, with longish hair that flopped engagingly over his forehead. His kindly face belied the strength of character that lay behind it. All three of his marriages ended in failure, I suspect because of the priority he universally gave to his students. Many indeed were the hours he spent, both day and night, interacting with them in seminars in the Department or informal chats in the coffee shop across the road from North Gate, where he would sketch thermodynamic phase diagrams on paper napkins.

The Rotary Foundation Fellowship I held had a single academic year tenure, which in my case was 1966-67; after that I was expected to return to Australia and speak to Rotary Clubs at home about my American experiences. As that year progressed I realised that it was here, at Berkeley, that I should undertake the PhD that had been lurking as an ambition in the hidden spaces of my mind since completion of my Sydney University Honours degree at the end of 1964. I was challenged by and at times intimidated by Ian Carmichael but in him, and in that inspiring intellectual environment, I recognised a once-in-a-lifetime opportunity to scale a peak that I had thought would never be nearby, let alone climbable. A fierce determination to complete a PhD in Petrology at the University of California, Berkeley, under Ian Carmichael as supervisor, invaded my soul and I would hear no argument against it, notwithstanding the complications it would entail. And complications there were, in abundance.

The first complication was that I had to go back to Australia at the end of the 1966-67 academic year. Getting home would be no problem; thanks to Rotary I already had my air ticket. Once home, I would have to address at least a few Rotary Clubs to fulfil my obligations to the Rotary Foundation, so returning to Berkeley immediately would not be possible. Further complicating my plans was my impending nuptials – Margaret and I were due to be married just a month after my return to Sydney. I would have to find a way to fund us both back to Berkeley and then work out how we would support ourselves once there. It was not as though we had much in the way of accumulated savings or had wealthy families behind us. I expected that Margaret would find a job when we finally got to Berkeley but knew that could take some time, even with her rapidly developing data processing (IT) skills. As a non-permanent resident of California, I would have to pay substantial tuition fees, unless I could swing a fee waiver. Even assuming all these issues could be resolved, there remained a major hurdle to achieving this PhD goal — what would I study for my dissertation?

3

THANK YOU, MR FORD

As my Rotary Foundation Fellowship year rolled forward at a rapid pace I made sure that I took every opportunity I could to use my travel allowance and see something of Western America. In those days, the Greyhound Bus Company was offering a "99 Days for $99" ticket, which was valid for travel anywhere on the Greyhound network, including both the USA and Canada, on the condition that there was no back tracking and only a single departure from the point of origin. The university year consisted of three terms, with a break of two weeks or so between terms. It became apparent to me that, with careful planning, I could set off on a trip at the Christmas break and travel for two weeks, then continue for another week or so within the 99-day period at the next term break. So off I went, starting from Oakland (near Berkeley) on Greyhound Bus No. 8118, travelling to Los Angeles and Disneyland, then on to the Grand Canyon in Arizona, to Salt Lake City and the Mormon Tabernacle in Utah, to Calgary in Alberta, Canada, across the magnificent Canadian Rockies, in winter, to Vancouver, where I knew some people to stay with and celebrate Christmas. After that it was back to California, not to Oakland but to Sacramento, where I bought another ticket for the short trip to Oakland and Berkeley. At the next term break I reversed the process, with a ticket back to Sacramento, where I resumed my 99 Days for $99, travelling to Las Vegas and Denver, finally finishing back in Oakland on Day 99.

The Greyhound bus trips were a great joy and my eyes were constantly agog at the superb and diverse scenery of western North America, to say nothing of the bright lights of Las Vegas! Disneyland lived up to my expectations, even if I was mildly rebuked as I sampled its delights alone: "Why so serious? This is the happiest place in the world," said the ticket seller, looking at my rather impassive face as I queued for yet another ride.

Trekking to the bottom of the Grand Canyon on the back of a mule, in freezing December temperatures, was an experience neither I nor my backside will ever forget. It was a giant geology text book laid open at all the most interesting parts. So immense is the canyon, and so deep, that, in modern parlance, it seems easier to conceive of it as a digital image produced for an epic movie than as a reality of nature. At first sight, it looks even more extreme than the CGI created for *"The Lord of the Rings"* or *"Star Wars"*. Not that I thought in those terms in the pre-digital age, when it was still true that the camera did not lie. But for me, both then and now, the key to the Grand Canyon's impact, as I recall standing at Bright Angel Point and absorbing the atmospherics of the place, is its graphic demonstration that the earth truly has depth, as well as length and breadth. The countless layers of sedimentary rock that are stacked upon each other from the bottom of the canyon, 5,000 feet below, to the very rim on which I stood, testify to the extent of earth history and the vastness of time that allowed such accumulations of sand to take place. When you see the Grand Canyon in the flesh, so to speak, you appreciate why it takes four dimensions – length, breadth, depth and time – to really understand the planet we live on.

After the Grand Canyon, my next stop was Salt Lake City in Utah. This required a connection in Flagstaff, Arizona, where my $99 ticket would put me on a Continental Trailways bus because Greyhound did not service that route. I can still hear the earnest young man behind the counter of the Continental Trailways waiting room in Flagstaff, as he answered the telephone:

"Good afffternooooon, Cont-unnn-ent-al Trayal-wayeees" he would drawl, in an accent that sounded like a Hollywood caricature.

Before my visit to Salt Lake City I knew only a little about Mormonism and that little was not very favourable. After watching an informative film in the Visitor Center at the Tabernacle I knew quite a bit more about the Latter Day Saints but my view of the religion did not improve. I liked their organ though, originally built in 1867 by Englishman Joseph Ridges and re-built in 1948, it contains 11,623 pipes, 147 voices (tone colours) and 206 ranks (rows of pipes) and produces a wonderful sound. Combined with the famous Mormon Tabernacle Choir it makes an impact that is justifiably famous.

Calgary in December was cold and I did not linger, boarding yet another Greyhound bus for a day and overnight journey to Vancouver. Crossing the Canadian Rockies, even in early winter, was spectacular, especially for a young Aussie who had hardly ever seen snow before that time, or, for that matter, real mountains. The Rocky Mountains are aptly named and when draped in snow they form breathtaking vistas. In some respects, they are the antithesis of the Grand Canyon, expressing the third dimension as height rather than depth but no less impressively. In the famous view across Lake Louise at Banff the layers of earth history rise dramatically upwards. Rocks formed beneath the sea stare down from the heights, creating not just one of the most stunning views on earth but also demonstrating very graphically the dynamic nature of Mother Earth. I was simply spellbound.

Christmas in Vancouver with people I had met on board the S.S. *"Orsova" en route* to the USA the previous September was very pleasant and they took great pride in showing me their beautiful city, including Stanley Park, with its carved and painted Indian totem poles.

The Christmas – New Year Greyhound bus trip, which covered about 6,000 miles, seriously, agonisingly, whetted my appetite for this vast and varied continent and I wanted more. As I settled back into the discipline of study in Term 2, I began to think seriously about realising my PhD ambitions by returning to Berkeley with my bride-to-be. But the barriers to that objective had not receded with time. There was still the issue of money, or more correctly, the lack of it, and as yet no research topic, although I was pretty confident that, under Ian Carmichael, a suitable topic would emerge in time.

Sometimes what appears to be an intractable set of problems can suddenly dissolve into thin air like a dispersing mist. At other times, the complications of life become so overwhelming they seem like a dense fog that refuses to lift and reveal a way forward. I feared a systemic collapse as a real 'pea souper' began to envelop me. My dreams of completing a Berkeley PhD were rapidly falling into an impecunious heap. Departure from Berkeley was locked in for early June so there were only a few months left in which to resolve this impasse.

Blessedly for me, and subsequently for Margaret, the dispersing fog turned out to be the more apposite metaphor as the academic year quickly swallowed the first months of 1967 and the June deadline drew ever nearer. The radiance dispersing that fog came in the form of a man called Lindley Sale. I had met Lindley and his wife, Peg, quite early in my tenure at Berkeley, as they were neighbours of my host at the Berkeley Rotary Club, Clyde Wilson-Reid. The Sales were a friendly, extremely hospitable family, who welcomed me and my Canadian flatmate, Maurice Manel, into their home regularly for Sunday evening home-cooked dinners. These were always followed by a lively discussion. Lindley and Peg were about the same age as my parents and they lived quite close to the university, where Lindley worked in the Administration. They were very kind to me and in addition to the much-appreciated dinners they invited me to join them on two occasions when they visited relatives on a ranch near Red Bluff in northern California. Their great hospitality continued unabated when, after our stint in Papua New Guinea, I returned to Berkeley with Margaret, my new wife. As their kindness was lavished upon us both, they became our *de facto* American parents. Meeting the Sales and some of their relatives and sharing those times with them allowed me to delve far more deeply into Middle America than would otherwise have been possible in the rarefied atmosphere of a university campus. Or under the superficial bonhomie of a Rotary Club. It was also the key to unlocking my dilemma over how to carry on at Berkeley.

Lindley Sale was the quintessential Quiet American: Tall, balding, genial and generous, with smooth and somewhat shiny skin that seemed immune to the California sun. He was always well dressed and usually sported a bow tie, which sat well on him and suited his style. He could have played the lead in *"Father Knows Best"*, the 1950s television sitcom that starred Robert Young. Lindley was also a fourth generation Californian, descended from a tinker who sold pots and pans to the miners during the gold rush to the Californian Mother Lode country, in the foothills of the Sierra Nevada Mountains. That gold rush began in 1849 (which is why the San Francisco football team is called "The 49'ers") and there was obviously money in pots and pans as Lindley's forbears established what later became the Safeway Supermarket chain. He was insatiably curious about Australia and was the perfect host, making me feel important by pumping me with endless questions about my native land. Like many Americans of his generation, Lindley had not travelled overseas much, except to Europe as a soldier during World War II, when he served in command with the Nisei forces (Japanese Americans, mainly from Hawaii). Nor did Lindley express much desire to visit Europe as a tourist in the 1960s. He did, however, confess to an abiding interest in visiting Australia, something that he and Peg subsequently did a couple of times after we were back in Sydney ourselves.

Lindley's financial generosity was displayed often, as many were the times we were treated to a meal at *Spengers* or another of Berkeley's restaurants. The huge *"Spengers Fish Grotto"* was very much an institution in Berkeley and still is; it was certainly Lindley's favourite. Apart from superb sea food, they offered each dining table a whole loaf of the famous San Francisco sour dough bread, which in those days I could eat without getting indigestion! The restaurant did not take reservations and diners had to wait in the bar area till called to a table:

"Sale, party of four," would come over the PA system after what was sometimes a long wait. Lindley, ever resourceful, eventually took to entering via the back door and slipping a $20 bill to a particular waiter, who then seemed able to find us a table remarkably quickly.

But it was Lindley's graciousness and generosity of spirit that really characterised the man and served as a model for me. To Margaret and me, Lindley and Peg represented all that was good about America at a time when so many were critical of its culture. Lindley, like Ian Carmichael, is a hero of my past and a man who, quite differently from Ian but just as assuredly, helped shape my future and make me into the person I am today.

When Margaret and I, in our turn, hosted overseas students visiting Australia we remembered well what a difference Lindley and

Peg had made to our sojourn in California. We vowed to follow their example. I am happy to say that we still to this day keep in touch with two such overseas visitors (and their parents), Hilary Johnn, a young woman from England who we see quite often, and Martin Sloboda, a young man from Slovakia who we see occasionally; our lives have been significantly enriched by that experience.

Lindley worked in the Office of the Chancellor of U C Berkeley (the administrative and academic head of the Berkeley campus, equivalent to an Australian university Vice-Chancellor). It was never made all that clear to me just what Lindley did there, and silly me never thought to ask, but I understood it to be a senior role. One Sunday evening, I think it was in March, 1967, he made a revelation that was to change my life. For some weeks, I had been telling him of my desire to continue studying at Berkeley and to complete a PhD, but lamenting that I could not see how it would be possible. My anguish was made all the more palpable by the fact that, after talking it over with Ian Carmichael, I had settled on a study of active and dormant volcanoes in Papua New Guinea as the proposed topic for my PhD research.

I had been in touch with Dr Norm Fisher, a renowned geological veteran of Papua New Guinea, who was in 1967 Director of the Bureau of Mineral Resources, Geology and Geophysics in Canberra. Fisher had suggested that I focus on the Willaumez Peninsula of New Britain, centred on a tiny settlement called Talasea. This locality, Fisher believed, would be suitable for study and physically amenable to field work because it was a long, narrow peninsula that was accessible by boat. It was a geologically pristine area, completely unmapped and unstudied, although obviously volcanic, with several prominent volcanoes and a large, well defined caldera. But my ambition remained totally unfunded and although Carmichael was sympathetic and wanted me as a full member of his graduate student clutch, he was still a relatively junior professor and could offer me no financial help. Then Lindley made his revelation and everything changed.

"Well now Garry, I've heard about a fund administered by the Chancellor's Office that might help you carry on at Berkeley," he said, settling back after dinner into his favourite arm chair, raising the footrest and lighting yet another cigarette (he eventually died of heart disease).

"It's been endowed by the Ford Foundation and has very general terms of reference, something like *for the improvement of graduate education*'."

"Why don't you apply for a grant from this fund to assist with your work in New Guinea?"

Neither Ian Carmichael nor anyone else in the Department of Geology and Geophysics was aware of this fund and Ian, though supportive, was quite sceptical about my chances of success. Undeterred, I lodged an application, requesting sufficient funding to cover both the cost of three month's field work in PNG and the cost of my return to California afterwards.

Much to the dismay, even chagrin, of my American fellow graduate students, and the Department's Geology professors, I was successful. Carmichael was stunned:

"Jesus Christ, Lowder, how the fuck did you manage that?"

"Goddamned Aussies," grumbled some of my peers, "what's your secret?"

"The quality of the application and the merit of my proposal, of course," I answered, tongue firmly in cheek. I did not want to draw attention to my contact in the Chancellor's office.

The amount awarded was $4,300, which does not seem like much today and indeed, was not quite enough, although they granted me another $1,000 once the work was underway. The total grant of $5,300 proved to be just sufficient to meet my needs in the cost structure of those years before the 1970's inflationary explosion.

To this day, I do not know what role Lindley played behind the scenes in the Chancellor's Office, although I am sure he was a strong advocate on my behalf at least. In any event, my reputation in the Department soared as the fog over the Bay cleared into a bright sunny day.

4

GOLDEN GATE

It was a cold and windy morning on the 16th of December, 1967, when, in pre-dawn darkness, I went out on a forward deck of the S.S. *"Oriana"* to watch the Golden Gate Bridge pass over my head. The wind whipped over my face as, shoulders hunched, I pulled my jacket tight around me. I stood alone; no one else was foolish enough to stand out there in those conditions. It was their loss, though, as the lights of San Francisco twinkled a warm welcome that belied the chill imposed by the elements. My pulse quickened with the emotion of the moment. Fear of an uncertain future was diluted by elation: I was finally back in California, together with Margaret, my bride of five months. Slowly the *"Oriana"* approached her berth in the Port of San Francisco. Docking was delayed by the strong winds, but eventually, lit by an insipid sun rising from behind the East Bay, the vessel was alongside the Embarcadero wharf.

The voyage from Sydney had taken about three weeks, with calls at Suva, Honolulu and Vancouver and we regarded that time as a delayed honeymoon, having gone to Papua New Guinea just days after we were married. It had been a good cruise, allowing us time to adjust to each other in a way impossible while living in a *Haus Kiap*[3] in PNG, where little brown eyes could often be seen staring upwards through the bamboo slats that served as flooring. We had even been

---

[3] A native village house preserved for visiting government officials ("Kiaps") and other VIPs.

successful in one of the social highlights of the cruise, when Margaret, as jockey, and I, as trainer, had won the "Horse Racing" competition.

For those who have not encountered this quaint form of shipboard entertainment that was very popular on passenger vessels in pre-digital days, let me explain. A structure is erected in the Ballroom that contains about a dozen tracks along which the "horses" are drawn when the "jockeys" wind the strings to which the horses are attached on to hand-held spindles at the other end. I had run into the game just over a year earlier when I first went to California aboard the *"Orsova"* and had learned then the secret of success at this caper. The spindle on to which the string is wound is not cylindrical, it tapers at both ends, which makes the string prone to slip off to one side, substantially reducing the length of string wound on for each revolution of the handle. The trick, as I showed Margaret during "track work" earlier in the day, is to turn the spindle through 45° such that the wound string does not slip to the narrow edge but follows a longer, elliptical path, generating much more retrieval for each revolution. She practised this manoeuvre patiently during the day and became adept at maximising the value of each turn of the handle.

Applying this new found skill, Margaret won her heat comfortably and recorded the fastest time of all the heats, by a considerable margin. The ship's tote made her a strong favourite for the final as I listened to the track-side chatter:

"It was just a fluke."

"There's no way she can repeat that time."

"She didn't seem to be winding any faster than the others."

"I don't know how she did it but I'm backing her to do it again."

No one seemed to notice that Margaret had changed the angle of the spindle as she wound.

The pressure was on as the horses lined up for the final.

"They're off!"

The race caller added tension and excitement to the atmosphere as the twelve heat-winners, all ladies, wound furiously to draw their horses down the tracks. Margaret, a picture of self-assurance, subtly turned her spindle through 45° and rotated the handle steadily, with confidence and just the glimmer of a smile on her face. She knew she had the race under control. Loud cheers erupted as her horse crossed the finish line, once again well ahead of her rivals, in a time that was even better than that recorded in the heat.

My dear wife, at the tender age of 22, was the toast of the punters that night, a celebrity without peer. It all seems a bit trivial now, but of all our many moments of trivia in the past 50 years, that night aboard the *"Oriana"* remains one of the most unforgettable. We still recall that night with good humour and a wry smile when we look at the official photo from the occasion. It shows the ship's Captain and Commodore of the P & O Fleet handing us the champagne prize, all dressed up in his white dinner jacket and black bow tie, and proudly displaying a "medal" on his trouser fly.

But, as they say, all good things must come to an end. The voyage was over; it was time to get serious. We now faced the harsh realities of settling in to a foreign land and justifying the confidence that had been placed in me by the University, by Ian Carmichael, by my parents and especially by my In-laws, Jack and Marie Freudenstein. They had been very supportive as their new but impecunious son-in-law had taken their beloved daughter off to the other side of the world with just the $1,000 that Margaret had saved standing between us and poverty.

***

The Commodore of the P & O Fleet celebrates with the jockey and trainer in 1967

Lindley Sale was dockside to meet us and take us to his home in Kensington, alongside Berkeley, where we would stay over Christmas. Later, early in the New Year, we moved into "University Village", an apartment complex for students in nearby Albany, where we had managed to procure an apartment. It was owned by the university, which also provided a bus service to the campus.

The apartment in University Village was small and sparsely furnished but comfortable enough and we were soon supplied with most of our kitchen utensil needs by Peg Sale, who took delight in mothering us. Other essentials, such as sheets and blankets were purchased at the J C Penney discount store, using as little as possible of our $1,000 in capital. As Second Term began I was expecting to make an immediate start in a paid role as a Teaching Assistant in the Geology Department, making a modest income but sufficient for survival until Margaret was working.

On the first day of the new term I approached the Department's

Administration Office confidently, only to be totally deflated when told by the Secretary:

"We thought you were not coming back till next term. We don't have a position for you till then."

"Bugger!" I said to Margaret that night, "How in the hell are we going to survive with no income for another three months?"

The pulse-quickening fear of failure loomed large once again, as the $1,000 sum shrank to $750, then to $500; we were living cheaply, and the Sales, bless them, were feeding us regularly. But Margaret had to get a job and fast. She had skills that were in demand (they would be called 'IT' today) but she did not have the "Green Card" that would allow her to work in America. The procedure we had been advised to follow was to get a job offer and then have the potential employer petition the Government for her to get an immigrant visa (shades of 457 Visas in Australia). She would then be entitled to a "Green Card".

After an exhaustive but relatively brief search, that is exactly what happened. Margaret was offered a job with Blue Cross, the major health insurer in Northern California, based in nearby Oakland. It was a good job, well paid and with interesting workmates. By the end of February, 1968, life began to take on some semblance of normality as I happily made the transition from impecunious student with dependent wife to impecunious student with rich wife; well, relatively speaking anyway. Margaret commuted by bus from Albany to Oakland for work; I commuted from Albany to Berkeley in the University bus to study. We made friends with Dick and Sally Williams, who lived in the apartment above us and the Williams remain friends to this day. Dick was undertaking a PhD in microbiology and later became a renowned and highly successful geneticist.

Margaret's job at Blue Cross soon put us on our feet financially.

Over the next three years her substantial and growing income enabled us to engage fully with the University community and to sample the sights and sounds of America's favourite city – San Francisco. It also provided the means for us to travel and explore many of the geographic wonders of Western America, especially the magnificent High Sierra. Our first taste of what this famous mountain range had to offer had come even before Blue Cross appeared on the scene.

It began late on a cool but clear and sunny afternoon, typical of Berkeley in the winter, when I was asked to report to the office of my supervisor, Ian Carmichael. As I stepped out of the elevator on Level 4 of the University's Earth Sciences Building my eye was drawn to the view visible through the glass doors that led out onto the balcony at the front.

"Hmmm, not bad at all," I muttered to myself, as I gazed across the Bay at the Golden Gate Bridge, its red paint glowing fiercely in the lowering sun.

"I don't think I'll ever get used to it."

I never did.

It was not easy to drag myself away from such a scene but my summons from on high could not be ignored, so I continued around the corner to Room 419 and entered Carmichael's office.

"You wanted to see me?"

"Yes my lad, I've got a job for you. I need some samples of obsidian from the Mono Craters for my research."

"You know where I mean," he continued, "we went there on an excursion last year."

"Sure", I replied confidently, "just to the south of Mono Lake on the other side of the Sierra. I remember it well."

"I want you to get up there and collect the samples for me. You can use one of the University's pool vehicles on my research account. Take Margaret with you and make a weekend of it."

It was only a week or so into the new year (1968); we were still settling in and Margaret was busy looking for work but this was an offer I could not refuse.

My first exposure to recent lava at Mt Lassen in northern California

5

BLACK GLASS

WATER

On the eastern side of California's Sierra Nevada Mountains, at an altitude of around 6,400 feet, there is a remarkable body of water known as Mono Lake. It is a hot, dry environment in the summer and a cold, often snowy place in the winter. The lake covers roughly 100 square miles and includes a prominent island – Paoha Island – formed by a volcanic eruption about 350 years ago. The stark scenery is popular with visitors, especially those with a geological or ecological bent, as there is much evidence of very recent volcanic activity in the area and the lake itself hosts a distinctive fauna, including brine shrimp and alkali flies, as well as the birds that feed on them. The stark, high desert scenery at Mono Lake has inspired both naturalists and film-makers for decades. Clint Eastwood's classic "High Plains Drifter", made in 1973, is the best known example of a movie shot there. Mono Lake is also where, in 1967, my wife and I came as close to death by exposure as we would ever care to.

Ian Carmichael had discovered that each occurrence of obsidian (black volcanic glass) in the western United States had a unique assemblage of trace elements. He analysed (non-destructively, using x-ray fluorescence) the trace elements in Indian arrow heads made from obsidian that had been collected as artefacts all over Western

America. From this he found that it was possible to track them back to their source and thus map out old trade routes used by the native people.

Ian had met Margaret when he came to review my field work in Papua New Guinea and the new bride had acquitted herself well with the visiting VIP. I guessed that Ian did not really need samples of obsidian from the Mono Craters quite so urgently but instead had an ulterior motive. It was the first time, but far from the last time, that this complex and challenging man, who became my greatest mentor, showed me his soft and generous side – he was giving me the chance to introduce Margaret to the magnificent High Sierra at his expense.

***

The vehicle was a large Ford station wagon, typical of American cars at the time but bigger than anything I had driven before. Its size would enable us to camp in the back of it overnight when we reached Mono Lake, which was important as we were travelling on a very limited budget. We knew it would be cold so we had borrowed warm sleeping bags and assembled as much warm clothing as we owned at the time.

Setting out early on a sunny Friday morning, the drive took us along the Interstate 80 Freeway, through Sacramento, the state capital, and on up into the Sierra Nevada Mountains. The road rises steadily until it crosses Donner Pass, at an altitude of 7,200 feet. There was a lot of snow around us as we traversed the pass but the road was clear. Descending from the mountains we crossed the state line into Nevada and drove on till we entered *"The Biggest Little City in the World"*, as the gambling city of Reno proudly proclaims itself. It was our intention to camp overnight in the back of the big Ford wagon when we got to Mono Lake, but cheap motels are legion in Reno so for this first night we indulged ourselves with a $6 motel room. Margaret was incredulous that evening as we strolled through some of the casinos, where grim-faced gamblers were pouring money into row after row of slot machines.

"I can't believe these people," she said, "why do they do it?"

"Just listen," I replied, as here and there brief bursts of *'ratta-tat-tat'* were heard as coins won by a few lucky punters dropped out of the machines.

"That sound is addictive; they can't help themselves."

***

The next morning, we headed south under an overcast sky on Highway US395, past the Nevada capital, Carson City, and continued on parallel to the eastern flank of the snow-capped Sierra Nevada Mountains. After a couple of hours driving the highway crossed back into California at Topaz Lake. Ultimately, US395 leads down to Los Angeles but our destination was much closer at hand, as less than two hours later we pulled into the hamlet of Lee Vining. After acquiring coffee and a snack for lunch at the truck stop, and topping up the fuel tank, we were soon back on US395 and drove south for another 5 miles or so. The road took us past the western shore of Mono Lake, to the junction with State Highway 120, which heads east, linking with US6 near the Nevada border. My intention was to follow this road as I knew it would take us across the line of volcanic domes and cinder cones that constitute the Mono Craters. Our objective, Panum Crater, dominates the Mono Lake skyline so we could not miss it. But as we turned onto Highway 120 my well laid plans struck a roadblock, literally.

"Garry, the sign says the road is closed."

Margaret stood before me under a leaden sky, with her hand on the 'ROAD CLOSED' barrier, which partly blocked vehicular access. She looked very dubious as I responded:

"No worries, we are only going in about four miles. Further east the road goes up over a range where it'll be snowed in; that's why it's closed."

"Besides," I continued, "look around you. There's no snow at all here and it's a flat drive to where we'll park. I can get around the barrier and it'll save us a lot of walking."

Having got us this far without incident, I allowed youthful exuberance to override common sense in my intention to get us as close as possible to Panum Crater.

***

It was exactly as I had said – we drove east along Highway 120 for four miles and parked on the road, facing back to the west on a gentle slope, where the road passed between two craters. Our geological objective, Panum Crater, was situated about half way between the road and the lake to our north.

"Before we go geologising, let's take a look at the lake itself," I proposed.

"We can follow this trail here." I pointed to a marked trail head beside the road.

"It'll take us to the lake shore where we can have a look at some of the tufa towers this place is famous for. But it's pretty damn cold, so we'd better rug up as much as we can."

Wise words, as it turned out.

***

The trail to the lake shore was well marked and it was a pleasant walk of a mile or so in the brisk air. Information signs erected along the way described the landscape we were crossing and its local geology and ecology.

"You'll find they do this sort of thing really well in America," I remarked to Margaret, as we crunched along the gravelly path towards the lake.

"Pompous ass!" Could that really be what she said in reply?

Mono Lake is indeed particularly renowned for the tufa towers that rise from the lake surface and impart a graphic, "other-worldly" appearance to the landscape. A modern drop in the level of the lake, due mainly to an insatiable demand for water by the city of Los Angeles, which diverted streams flowing into Mono Lake in the 1940s, has exposed a great many tufa towers to the air. A public campaign, culminating in a court case in 1983, forced Los Angeles to reduce its off-take from the Mono Lake catchment and the lake has started to recover, although it will be decades until the lake achieves the level it was at before the incursion by the thirsty southerners.

Upon reaching the lake we settled down on a convenient rock outcrop and munched on our snacks as I explained the origin of the tufa towers to Margaret:

"They form under water by the precipitation of calcium carbonate where calcium-laden springs seep out of the lake floor and react with carbonate dissolved in the lake water, gradually building mounds that become towers as much as 30 feet high."

"The process is a bit like the formation of stalactites in limestone caves, but in reverse."

Margaret absorbed this information as she stood up and strolled along the waterfront.

"They look like sentries standing guard," she called, gazing at the tufa towers nearby, "and I think I can see some of those brine shrimp in the water but where are the flies and birds you told me about?"

"Too cold," I replied, "you'd have to come in the summer to see them."

***

After an hour or so prowling around the Mono Lake shoreline I decided we had better turn our attention to the reason we were there.

"It's not getting any warmer," I noted, as we trekked cross-country to Panum Crater, where a black and glassy obsidian lava flow contrasted strongly with the grey landscape nearby. It was obviously a very recent feature geologically. Upon arriving at the edge of the flow, Margaret was quick to observe:

"Didn't I see something like this at Talasea?"[4]

"You did, there was a big outcrop of it on the road to Volupai."

"I can see why they used obsidian for making arrow heads," Margaret continued, as I began chipping at the rock.

"It seems to want to break into curved faces with very sharp edges."

"It does, and those surfaces are called conchoidal fracture."

"Why is it glass instead of normal lava?" she asked.

"Obsidian is very high in silica, about the same as window glass. At depth, it also contains quite a bit of dissolved water, which helps to keep it fluid at moderate temperatures. Once it gets to the surface the water boils off and the viscosity of the lava goes way up."

"It quickly becomes too sticky to crystallise properly and once the temperature drops a bit it forms glass rather than crystalline rock like normal lava."

I continued with my sampling while Margaret wandered around

---

[4] See Chapter 15.

looking for evidence of a pre-historic workshop. Each sample was given a number and a description of its field setting was recorded into my yellow-covered notebook. Anxious to impress my Prof, I focussed intently on the task at hand, not noticing the gathering gloom. Finally, I was done and after putting the last of a dozen or so specimens into my backpack, I stood up and arched my back into a good stretch. As I did so my eyes were drawn to the overcast sky, which I could now see was much darker and more threatening than it had been when we left the vehicle a few hours earlier. I was still confident but my initial bravado was now tempered by a pang of doubt, which only increased as I felt the cold sting of the first snowflakes on my face.

"Shit, it's beginning to snow. We'd better get out of here," I declared.

We quickly gathered up our things and set off through the sagebrush back towards the car, moving rather more rapidly than we had on the outward journey, notwithstanding the load of rocks in my backpack. There was little conversation as we retreated, with both of us feeling increasingly apprehensive. It was remarkable and quite frightening how quickly the bleached grey landscape was turning white.

"I don't think I have enough clothes on," said Margaret.

"Hmmm. We'll be right, just walk faster," I replied, trying to exude confidence, which was visibly betrayed by my uncontrolled shivering. Was it the cold or fear?

Snow was beginning to accumulate on the road and I was pleased we had left the vehicle facing back to where we needed to go. Piling our gear, including the precious rock samples, into the back of the Ford, we jumped into the front and I turned on the ignition. The starter motor growled, and growled, and growled, but the engine did not start. I tried again, to no avail. Fearful of flattening the battery, I suggested that we had better try a rolling start, not knowing then that this is impossible with an automatic transmission. The Ford rolled down the gentle slope and came to a silent stop. No hint of an engine start. Laboriously, we tried pushing the car back up the hill to try again, but it was futile.

"Bloody hell, now what do we do?"

"We're going to have to walk out to the main road to get help, of course," Margaret replied tersely, venting the rising tension between us.

"It'll be a bit of a challenge to walk four miles in these conditions," I declared, stating the bleeding obvious, "but we're on a closed road. No one knows we're here and no one will come to our rescue if we stay. I guess we don't have any choice."

The snow was falling heavier than ever, the temperature seemed to be dropping even more and now the wind had come up. We were in a blizzard and in more danger than we either realised or wanted to admit.

"Well, come on then, we'd better get going." I tried to sound reassuring to Margaret while inwardly feeling foolish to have put us in this predicament. Suddenly I felt very cold and quite scared.

***

But we were both young, fit and quite used to walking so we struck out energetically towards the main highway. It was good to be moving and it gave us something to focus on rather than my naïve bravado. We had dressed as well as we could for the cold but it was not enough for a blizzard. As we trudged along the road, now a smooth white pathway, our ears, noses and finger tips grew numb, for we had no hats or gloves. Each exhalation of breath became an icy fog. The wind whipped and eddied around us, making the snowflakes feel like shards of glass as they struck our faces. Apart from the wind-driven snow, everything else was still and quiet as the landscape of rock and sagebrush disappeared beneath a blanket of white.

It is remarkable how far a simple four miles can be when the circumstances are dire. After an hour or so, with little idea of how far we had come, we took a breather. Our clothes had become icy carapaces; icicles hung from our eyebrows as we squinted against the glare; my three-day old beard was caked in ice and our hair had become crisp, white helmets. But, though we were tiring, we resumed the trek rather than rest longer and succumb to the cold. Encouraging each other vocally to hide our concern, we even laughed at the contrast between Mono Lake and the steamy jungles of Talasea, in Papua New Guinea, that we had so recently left behind.

"Wouldn't the Talasea crowd be amazed to see us now?" I asked. "They'd think we're crazy."

"Garry, they thought we were crazy the whole time we were in Talasea."

That shut me up, as another hour passed.

"We look like characters from Dr Zhivago" Margaret declared, as we finally reached US395 and rested ruefully against the "ROAD CLOSED" barrier. We felt better to have made it thus far, but we were by no means safe.

"We've got to keep going to Lee Vining. If we stay still we'll end up with frostbite."

Resolutely, we turned north into the wind on a highway covered in snow, clean, unmarked snow, showing that few if any vehicles had passed since the snowfall had started.

***

About twenty minutes later Margaret suddenly stopped.

"What's that?" she asked, as we both turned around, looking out over our footprints that disappeared into the white haze.

Then we heard it again – a grinding noise that sounded like … like an engine! Hallelujah! It was coming from the south and as the sound grew louder it became obvious that a large vehicle was approaching. Our hopes of rescue rose as we peered through the snow to see what was coming. Abruptly, out of the whiteness came a large, red truck with a blade across the front – it was a snow plough, clearing the road. Waving frantically, we put ourselves in the path of the truck, wanting to make sure we were not missed in the gloom. The truck stopped – we must have been a bizarre sight for the driver but at least with the way we were jumping up and down there was little chance he would not see us. A quick explanation of our circumstances was sufficient for him to invite us into the cabin with him, as he was heading into Lee Vining and would take us there. The warmth of the cabin was just wonderful after several hours of bitter cold, and the sense of relief was overwhelming. Feeling my eyes moisten, I was unsure whether it was tears of relief or just the melting snow on my icy face. Either way, I quietly reflected on how lucky we were. I turned to our rescuer – a large, middle aged man with a face that looked as weather-beaten as we felt – to explain in more detail how we had come to be there. He listened attentively as he guided the plough between the orange-coloured poles beside the road.

"You folks aren't used to the mountains, are you?" he said, speaking kindly.

I acknowledged our novice status.

"The weather changes very quickly up here. You shouldn't have gone in there without telling someone. Never mind, you're OK now and I'll drop you off at the truck stop in Lee Vining. Ask for Billy, he'll get you going again, although if I were you I'd stay in Lee Vining tonight. This snow's not over yet."

Severely chastened, I thanked him for his help and promised to be less foolhardy in the future.

***

Billy did get us going again, remarkably quickly. He looked almost as young as us, in spite of his thin, wispy beard, but he radiated confidence; a true mountain man. We assumed he was a mechanic but did not like to ask as he seemed to know what he was doing and we were just so grateful for his help. Bundling us into his Chevy pick-up he drove back down US395 to the Highway 120 turn-off and then followed my ill-fated route towards the craters. In but a few minutes we were once again standing beside our cold and lonely Ford. Nearly a foot of snow had accumulated on it in our absence, making it look very forlorn and unloved.

"What ya gotta understand with a Ford," Billy explained, as we stood shivering in the blowing snow, "is that they ain't much good at high altitudes."

"If ya wanna start a cold Ford up here in the winter ya gotta open the choke."

He demonstrated by removing the air filter and poking his finger into the choke diaphragm.

"Give 'er a go now," called Billy, as he held the choke open.

After a couple of growls, and to my immense relief, the car started, which was just as well as it was still very cold and snowy. I had chains in the back of the Ford (standard University issue) but decided not to take the time then to put them on, as the light was already poor and dusk was not far off. We began our exodus, driving slowly along the snow-covered road.

"Take it very steady" Billy advised.

I hardly needed reminding of this, but with no significant inclines, we did manage to stay on the road, even if there was a bit of slipping and sliding along the way. Billy followed in the Chevy close behind us, which was both reassuring and a bit nerve-wracking. Once we reached US395 the going got easier because the snow plough had been busy clearing the road and we arrived in Lee Vining just as the light faded. The danger was over but not the embarrassment; our rescuer would not take any money from us as he pointed us to the Lee Vining Motel (all eight rooms of it) across the road.

"It's pretty darn basic but it's all there is in this 'ere town," he said, apologetically.

Poor as we were, camping out in the back of the Ford did not seem like such a good idea after a day like this. So, thanking him profusely for his help, we gratefully booked a room, a warm, dry room, for the night.

The next morning, we looked out on a blinding whiteness. The snowfall had ceased and it was bright and sunny. As the sun rose it grew warm enough for the snow on the roads to begin melting. We stayed another day in the district, bunking down in the Ford the next night and using my new found mechanical expertise to get the station wagon started in the morning.

***

Two days later I walked into Ian Carmichael's office and placed the obsidian samples down gently beside the microscope on his bench.

"I hope these are what you wanted," I declared.

"Looks good," he replied, as he examined our costly cargo "but how did you get on with the weather? Didn't it snow up there on the weekend?"

"Yeah, a bit; no real problems though," was my blasé response.

I was not about to tell the man who would control my destiny for the next three years the sorry tale of our chilly initiation to the High Sierra.

6

DRY FLY

After a shaky start to 1968 life in Berkeley began to resolve itself into a pattern. Which is not to say it was routine – nothing was routine in Berkeley – but a steady schedule emerged that allowed us to settle into the community and feel quite at home. We stayed in the University Village for about a year but then moved to a pleasant one-bedroom apartment in Berkeley, just two blocks up Euclid Avenue from the North Gate of the campus, on the corner of Le Conte Avenue. From 2525 Le Conte there was a bus to Oakland for Margaret to get to work easily and for me it was a simple, five minute stroll down Euclid to the Department of Geology and Geophysics. This made it easy for me to go back and forth to the campus during the day if need be and was especially appreciated when I moved on in my research to lengthy sessions, day and night, on the Department's electron microprobe. It also took me daily past La Vals, the home of my first and (in my mind at least) best-ever pizza.

It is hard to believe in these days of an enlightened Australian cuisine that embraces foods and flavours from every corner of the world, but in 1966 pizza was completely unknown to me and, I suspect, to most Australians. It was only a month or so after I first started at Berkeley when my culinary re-education began:

"Hey Garry, a bunch of us are off to La Vals for pizza and beer. Ya wanna join us?"

Frank Huffman, a tall, friendly man from southern California, was one of several graduate students with whom I shared Room 207 in the Department. Spanish student, Enrico Merino, who sported a magnificent handlebar moustache, was another.

"Sure."

The 'beer' part I understood and I did not want to sound too ignorant or naïve by asking what a 'pizza' was. So off I trotted, one of a group of five or six, out through the North Gate, across Hearst Avenue and up Euclid Avenue a few doors to an archway that led into a courtyard – the famous "La Vals". Frank did the ordering. The beer arrived and we began to drink eagerly, as only students can. Then a short while later out came a waiter carrying two large trays on which sat some sort of pie or tart; there seemed to be a lot of tomato sauce and melted cheese and those little circles looked like slices of salami. A wheeled implement with a sharp edge was produced and Enrico proceeded to roll it back and forth across the tray, cutting the product that sat upon it into segments. My colleagues each withdrew a piece from one of the trays and started to eat. I copied them, trying to appear confident, even nonchalant.

Pow! My digestive glands went into overdrive as my mouth flooded with saliva. It was not so much the tomato paste, or the stringy cheese, or even the pepperoni, although they were delicious. It was the oregano and other herbs and spices in the sauce that imparted a flavour sensation entirely novel to me. As I withdrew a second segment to eat I knew that I had found a new and exotic food universe. I have been a fan of oregano ever since.

"Toto, we're not in Kansas anymore!" Well, not in Australia anyway.

La Vals quickly became a familiar haunt – what better way could there be to spend a Friday afternoon than to eat plenty of pizza and wash it down with copious quantities of beer? Albeit thin and almost tasteless American beer.

Parties were a regular social event throughout that year and the three years that followed, after I had returned to Berkeley with Margaret as my wife. No doubt lots of pot was being smoked at the time and I remember seeing someone with what he told me was hash at one party. But I was never offered, and conservative old me never sought to try, marijuana or any other drug. There cannot be many former students of Berkeley in the 60s who can claim that rather dubious distinction. There was always plenty of booze to consume, however. Not just beer but California wine. We used to stop in at the Oak Barrel Winery on University Avenue on the way to a party and pick up our favourite:

"A flagon of Red Mountain Vin Rosé please."

At $1.19 a gallon it was an economical way to dull the senses and relax into party mood. And if red eyes and a mountainous hangover the next day were an invariable consequence, so what? We were Berkeley grad students and thus invincible!

On a couple of occasions Ian Carmichael hosted a Friday afternoon party in the Geology Department's library. They became quite rowdy and raucous affairs, undoubtedly as a result of the witch's brew he concocted as a welcoming cocktail. Never one to do things by half, he would start with a large glass laboratory vessel, essentially spherical and about half a metre in diameter, with openings in the top and a long glass spout for pouring. In would go various fruit juices and flavourings and then chunks of dry ice for cooling (Americans insist on their drinks being chilled; a glass of Coke is often much more a glass of ice, with a little of the dark sweet liquid poured over it.) Then finally in went the alcohol. Not whisky, nor brandy, not vodka nor gin, but litres of pure, analytical grade ethyl alcohol were brought down from his laboratory upstairs and added to the by-now fuming, bubbling mix. It was such a powerful concoction that most of us lost our inhibitions very quickly, or were laid out asleep after just two or three drinks.

Now I don't want you to think that life was all beer and skittles at Berkeley, or even vin rosé and ethyl alcohol. We were serious students who were there to study, to research and to write up our results for publication and compile them into a dissertation, a thesis. But before anyone was permitted to proceed to a PhD, it was necessary to pass an oral exam. Orals success allowed a student to aim for the highest prize, a doctorate from Berkeley; orals failure meant relegation to a Master's degree at best. It was a significant barrier and one that ensured that only those with real talent would be granted a Berkeley PhD. That sounds immodest on my part, given that I was successful, but it was very much a "sheep from goats" process and it is one of the reasons that Berkeley PhDs are so highly valued.

The oral exam was a daunting procedure. The panel consisted of three professors from my own Department of Geology and Geophysics, plus two from other departments. I was able to choose who they should be and my panel consisted of my supervisor, Ian Carmichael (petrologist), together with Frank Turner (petrologist) and Garniss Curtis (geochronologist) from Geology, plus a professor from Paleontology and another from Geography (I confess that I forget their names). In theory, the panel members could ask any question they liked about any subject at all, in order to gauge your reaction to unexpected circumstances. In practice, most questions related to the Earth Sciences, although not just to my specialty (petrology). My predecessor students had compiled an "Orals Book" in which they recorded as much as they could remember about their own orals exam, including tricky questions favoured by the various professors. Turner was infamous for asking the precise chemical formula for the mineral hornblende (a member of the amphibole group), which is:

$(Ca,Na)_{2-3}(Mg,Fe^{+2},Fe^{+3},Al)_5Si_6(Si,Al)_2O_{22}(OH)_2$

I learned this off by heart, but of course, he did not ask it of me; instead, he asked (among many other things):

"Garry, how many symphonies did Mozart write?"

My answer: "41" (I was on safe ground there).

When he finally said, "No more questions," I breathed a sigh of relief, but then Carmichael piped up:

"Okay Garry, what is the formula for riebeckite?"

I was stumped. It, too, is an amphibole and has a simpler formula ($[Na_2][Fe_3^{2+}Fe_2^{3+}]Si_8O_{22}(OH)_2$), but for the life of me, I could not remember it. I knew it contained sodium and iron, a fact that I proffered up, but beyond that I admitted defeat. It did not matter in the end – they much preferred an honest "I don't know" than any attempt to bluff or pretend.

The palaeontologist gave me a rock sample to look at:

"What can you tell me about the geological environment this sample might have come from?"

It was a piece of limestone with fossil shell fragments embedded in it. I suggested it was from a back-reef environment, where coral and shell fragments eroded from the front of the reef are washed behind by waves and are later cemented by chemically precipitated lime ($CaCO_3$). Most American geology students do not study palaeontology as undergraduates and might have struggled with this question, so my confident answer seemed to surprise my inquisitor and satisfy him completely.

"Thank you, Dr Jenkins." The shadow of a smile touched my lips as I paid a mental tribute to the man who had taught palaeontology to me in my undergraduate days at Sydney University.

The geography professor confessed to being out of his depth but asked a few token questions about Papua New Guinea, where I had done my field work; these were despatched fairly easily.

Finally, after about two hours, I was asked to leave the room for a few minutes. Upon my return, I was relieved to see a cloth-wrapped bottle on the table. Seeing that, I knew I had passed, for the "Orals Book" had made it very clear:

"When you see a bottle of "Dry Fly" sherry on the table you can relax – it is Curtis's signal that you have passed."

And thus a major hurdle had been overcome and I was free to pursue my PhD goal without constraint.

***

Data for my research were generated in three main ways:

First came petrography, in which thin sections[5] of the rocks I had collected in Papua New Guinea were examined under a petrographic microscope. As an undergraduate at Sydney University I had to make my own sections, but here at Berkeley I was able to put this into the hands of a technician, who did a far better job than I would have done. One by one, I studied the PNG samples using Ian Carmichael's superb Carl Zeiss petrographic microscope. This instrument was a joy to use; compared with the cheap Japanese microscopes of my undergraduate days, the Zeiss was like driving a Mercedes SL 500 rather than a Datsun 120Y. Peering through the microscope at each

---

[5] Slices of rock are mounted on glass microscope slides and ground down until only about 0.03mm thick; at this level, most rock-forming minerals are transparent and can be identified and studied under a microscope with transmitted light.

slide I could determine the component mineralogy, grain size and texture, which in turn allowed me to identify the precise rock type. They were all volcanic rocks but varied a lot in composition, which was important when I later developed my thesis about their origin, evolution and relationships.

Second came electron microprobe micro-analysis. The microprobe was a tool for analysing the chemical composition of individual mineral grains, or parts of gains. This was achieved by firing a beam of electrons at a thin section (in a vacuum inside the probe) causing the spot irradiated (a few microns across) to emit x-rays. Each element emits x-rays of different wave lengths, allowing analytical discrimination. Three detectors collected the x-rays, which were generated proportionally to the concentration of the elements being analysed. Many long sessions were spent, usually late into the night and early morning, analysing grain after grain, sample after sample. Each spot was analysed for two minutes, which left plenty of time for me to puff away on my "Sherlock Holmes" pipe! (I was nothing if not the quintessential grad student!) After each reading the numbers of x-rays recorded were automatically typed out on an electric typewriter, which was later superseded by a punch card machine. By comparing the x-ray numbers generated for my samples with the numbers generated periodically for a standard sample whose composition was known, it was possible to calculate the concentrations of the elements analysed at each point.

The third key source of data was chemical analysis of whole rock samples. The trace elements were analysed by the relatively simple x-ray fluorescence ("XRF") technique but for the major elements (silicon, aluminium, iron, magnesium, calcium, sodium and potassium plus a few other minor ones) Ian Carmichael insisted that we learn to carry out wet chemical analysis. This was a much more difficult and laborious process, but the results, if done properly, were far superior to those from XRF. Insistence on wet chemistry was typical of Carmichael's uncompromising standards.

AIR

As I enter Room 417 in the Earth Sciences Building I again encounter a vague, slightly acrid chemical smell whose source is hard to identify. You would think I should be used to it by now, regular visitor that I am, but it still surprises and confronts me. It is not as though the caps have been left off the acid bottles, or that someone was dissolving a noxious compound outside the fume cupboard. Everyone follows the rules here, working in mortal fear of the Prof, who sits and calls out instructions just through that door into the adjacent Room 419. His students have free reign in his lab, but he will severely censure anyone who does not meet his standards of good laboratory housekeeping and best practice.

"You bastards make sure you clean up my lab and put everything back where it belongs!"

The Prof's voice drifts through the doorway, at once both intimidating and reassuring: He is always there for us.

All of us students have encountered his wrath at some time or other, the weaker ones amongst us cringing under his acerbic tongue. It is only after some time that we realise it is a case of barking rather than biting. His brusque façade hides a generous and gracious spirit that wants to see his students excel. But he will brook no sloppiness in this lab! That is why I am confident that the smell I encounter in this room is not toxic and represents no more than the fragrant aftermath of frequent use of these facilities by the Prof's most committed students.

Familiarity breeds not contempt but confidence. I am at ease here, in this lab, with its black bench top and white ceramic sinks under curved stainless taps. I am comfortable amongst its Bunsen burners, its tall burettes, its stubby pipettes and the racks of Pyrex beakers. I like the neatness of the technicoloured jars of chemicals on the shelves, where the bright yellow lead chromate is my favourite. I am constantly amazed to have scales that weigh to the last milligram. I am careful using the little brush to clean the crucible before weighing, mindful that the brush has a strip of radioactive plutonium attached to the top to discharge any static electricity that could affect the balance. I am cautious but not afraid of the hydrofluoric acid, in plastic bottles because it would dissolve glass. Nor am I intimidated by the slightly sinister fume cupboard sibilantly sucking in air over in the corner.

Precision wet chemical analysis of silicate rocks, the Prof insists, is an essential skill to acquire if we really want to be at the cutting edge of our science.

"Can't we do this by X-ray fluorescence Prof?" we ask, knowing how much easier and quicker that technique is.

"No bloody way! If you guys want to have data you

can trust then it's wet chemistry or nothing."

The emphasis is on precision, as we scorn those who slop a bit of solution on the floor and announce:

"That looks like about a tenth of a percent."

Here in this lab we bitch, we grumble, we mutter amongst ourselves, but we learn the discipline of the analyst. Here in this lab we discover that you only overdraw a pipette full of nitric acid once. Here in this lab we know we must be extra careful in titrating with the solution used for analysing aluminium, which tastes, I can say from sour experience, like concentrated jalapeño chillies. Here in this lab, we make the slow metamorphosis from naive students to qualified scientists.

"I am the very model of a modern...
grad student!"

7

WRIGGLING RAINBOWS

California has long held a special status in American history and culture. The land was extensively settled by indigenous people for centuries before sporadic visits by Spanish mariners began in the late 16th century. The establishment of missions began about 100 years later. It was these missions that later gave their names to many of the modern day cities and towns of California; San Francisco, San José, Santa Barbara, San Mateo and San Diego among them. The area became part of Mexico in 1821, following the Mexican War of Independence. In 1848, it was ceded to the United States after the Mexican-American War and California was admitted as the 31st State of the Union on September 9, 1850. By then the great California Gold Rush had been under way for more than two years, peaking in 1849.

Both geographically and geologically, California is a land of extremes and that propensity for outlandishness is reflected in its culture. The state is home to both the highest point (Mt Whitney, 14,495ft) and the lowest point (Death Valley, -282ft) in the 48 contiguous states. The former can be seen from the latter. In the east, the high mountains of the Sierra Nevada[6] form a solid barrier that challenged many a wagon train of settlers in the 19th century. In the west the San Andreas Fault, source of the 1906 earthquake in San Francisco, is a constant reminder of the restlessness of this continental margin. In the south, severe deserts isolate the coastal oasis that is the

---

[6] Spanish for "Snowy Mountains"

Los Angeles Basin. In the north, the Cascade Range, which dominates the skyline in Washington and Oregon, extends into California, where young volcanoes such as Mt Shasta (14,179ft) and Mt Lassen (10,457ft) express the subterranean instability that is responsible for so much of Western America's superb scenery. Mt Lassen last erupted in 1915 and a visit there with Ian Carmichael early in my tenure at Berkeley gave me my first direct exposure to recently active volcanism.

The state also boasts the world's tallest trees (California Coastal Redwood – *Sequoia sempervirens* – up to 115m high), the world's largest trees (Giant Sequoia or Mountain Redwood – *Sequoiadendron giganteum* – up to 17m in diameter) and the world's oldest trees (Bristlecone pine – *Pinus longaeva*) the oldest of which is more than 5,000 years old.

Old trees it has in abundance but the essence of Californian culture is the uninhibited exuberance of youth. Within the United States California has a reputation as the home of the weird and exotic. This was never more so than in the 1960s, when I saw firsthand the development of "Flower Power" and the Hippie revolution. Such cultural phenomena often tend to gravitate to university environs and Berkeley was no exception. Long hair, beads, leather vests and floral headbands were a common sight, especially on Telegraph Avenue. At first, conservative students like me tended to look askance at the Hippies but I soon learned to take it all in my stride. Indeed, after a year or so in Berkeley nothing, absolutely nothing, surprised me!

But it was the Sierra Nevada that really captured the imagination of a young geologist from the lowlands (a.k.a. Australia). The High Sierra became for Margaret and me a place of pilgrimage, such was the power of nature so graphically displayed by its glaciated peaks and forested valleys. Each summer saw us exploring those mountains and in the summer of 1968 we spent every weekend camping somewhere in the majestic High Sierra.

A large pluton of granite, known as the Sierra Nevada Batholith[7] forms the backbone of the Sierra, capped in places by young volcanic rocks and draped on the west by sedimentary and metamorphic rocks of the "Mother Lode" gold belt. It is the granite that gives the High Sierra its striking appearance, best seen from high viewpoints. Vast panoramas of nearly bare, light grey rock, wiped clean and polished by overriding Ice Age glaciers, can be seen rolling away into the distance. Between the peaks the glaciers carved U-shaped valleys and built hanging waterfalls where ice tributaries once joined the main flows. In many places, loose rocks carried along in the base of the moving ice have scratched their initials into the bedrock. These features are best exemplified in that most famous of California's National Parks – Yosemite[8]. The stunning sheer cliffs of *Half Dome* (rising 1,444m above the valley floor) and *El Capitán* (900m) are ice sculptures *in extremis*.

While the highest parts of the Sierra are open range and above the tree line, the valleys and slopes are well forested, principally pines of many types, dominated by the ubiquitous Ponderosa Pine, with its distinctive platy bark that reminded me of crocodile skin. Our usual practice for a weekend in the High Sierra was to find a campground in one of the many National Forests and pitch our little tent beside one of the picnic tables that the Forest Service so kindly provided. In the evening, a candle inside a glass jar on the table would supply just enough light for cooking and eating, with passers-by remarking:

"Oh, that is so cute!"

"Thank you," we would reply politely.

Little did they realise that our relatively impoverished state

---

[7] A batholith is a large body of intrusive igneous rock that formed deep below the surface.
[8] Pronounced: Yo-sem-it-ee.

permitted nothing better. When sometimes we were forced to use larger campgrounds the peace and serenity of our surroundings were often shattered by the sudden eruption of a petrol-driven electric generator nearby. That experience motivated us to seek out the primitive campgrounds where we could to avoid the Winnebago set.

Hiking in the High Sierra is an absolute joy, a fact well recognised by John Muir, renowned Scottish-American naturalist and strong advocate for wilderness protection, who explored these mountains in the 19th and early 20th century. Muir founded the Sierra Club and is often referred to as the "Father of the National Parks". Trails led away from most campgrounds and typically would traverse a mixture of glaciated pavement, pine forest, where there was little if any understory, and the shores of small glacial lakes, known as tarns. Trout populated every lake and stream and fishing in the mountains was a delightful experience, enhanced occasionally, speaking personally, by the actual capture of a wriggling rainbow.

When we had more time than just a weekend we ventured further afield, into Arizona, Utah, Idaho and Colorado. Each of these states had superb scenery to offer and our visits were enriched by exposure to the Native American (Indian) culture and history, especially the Navajo and Hopi in Arizona and the cliff dwellings of Mesa Verde in Colorado. Everywhere we went the geology of Western America thrust itself upon us and the role of that geology in forming the stunning scenery surrounding us was evident. Idaho offered the very recent lava fields of Craters of the Moon National Monument. Arizona hosts the one and only Grand Canyon but we also loved visiting the extinct volcanic cone of San Francisco Mountain near Flagstaff. In Utah, the graphically cross-bedded sandstone cliffs of Zion National park contrasted with the oddly shaped limestone pillars[9]

---

[9] Called "hoodoos"

of Bryce Canyon. Utah is also home to the yellow bluffs of Big Rock Candy Mountain (yes, there really is such a place). Colorado is one huge geology tutorial, where the broad north to south expanse of the Rocky Mountains, with 56 peaks above 14,000ft in elevation[10], dominates the landscape. Wherever we went, there was no denying that the diverse scenery we were enjoying was merely the surface expression of deep-seated and highly active earth forces. Western America is truly a geological smorgasbord.

As I said to Margaret at the time, and time after time in one form or another,

"We can appreciate how all this splendour came to be, but I can't help wondering what the average Joe from Chicago makes of it all."

"He probably thinks it is right and proper that the richest and most powerful country on earth should also be the most beautiful."

I did not disagree with her reply, but added:

"The most annoying thing about American arrogance is that it is so often well founded!"

My research based on the Papua New Guinea field work and the rock samples I had collected there was progressing well and although a return trip to Talasea would have been helpful to resolve certain queries that had arisen, it was financially untenable to do so. I suppose I could have pushed out my work on Talasea and made it sufficient for the desired PhD, but taking on a second project, closer at hand, seemed like a better option. Especially as I was still relishing the grad student life at Berkeley and Margaret was going from strength to strength in the role she had at Blue Cross. We were by then starting to build up some savings that would be very useful when we returned to Australia. I confess, therefore, to having a certain pecuniary interest

---

[10] But not one of them higher than Mt Whitney in California.

in staying on at Berkeley so Margaret could continue working at Blue Cross.

In the spring of 1969, therefore, I found myself in the southwest corner of Utah, mapping and sampling a suite of lava flows that had erupted onto a high plateau and spilled over the escarpment into the valley below. The last eruptions had occurred about 1,000 years ago and thus, in a geological timeframe, were very recent. That part of Utah forms the eastern margin of what is known as the Great Basin, or Basin and Range Province. This geological province extends from western Nevada to western Utah and is characterised by repeated 'horst' and 'graben' structures, formed when this part of the crust was stretched by earth movements to the west. The stretching was accommodated by the formation of a series of faults, like alternate forward and backward slashes in a URL code. Between each forward and backward slash was an uplifted zone (a 'horst', expressed as a mountain range); between each backward and forward slash was a subsidence (a 'graben', expressed as a valley). Due to this unusual structure, driving across Nevada on Highway US50 today is very much an up and down and up and down and up and down affair.

In this district the city of St George, which lies in a graben, is the low point, at elevations between 2,000 and 3,000ft above sea level. The Markagunt Plateau, my main area of interest, is a horst to the east of St George and stands mostly around 10,000ft a.s.l., with the highest point being Brian Head, at 11,307ft. To the north of St George, and quite a bit higher than it, is the smaller town of Cedar City, which sits near the western edge of the plateau. Cedar City was my local supply base but for the duration of my field work I camped out in a National Forest campground up on the plateau.

As it was almost summertime, the daytime temperature varied from very hot (44°C) in St George, to rather cool (10°C) on the high plateau. Similarly, the ecosystem ranged from arid desert with saguaro cactus in the low elevations to lush mountain meadows and pine forests, criss-crossed by gurgling brooks, in the high parts. Altogether an agreeable place to spend some time outdoors and a great contrast to the steamy jungles of Papua New Guinea.

Sharon's Pool

WATER

It was about two in the afternoon as I stood on Brian Head, reflecting on my good fortune. At just over 11,300ft above sea level the peak is the highest point on the Markagunt Plateau, which stretched into the distance a few hundred feet below me. Both to the south and to the north I could see some of the recent lava flows that were the reason I was there. Bare, roughly textured ribbons of black rock that reflected little of the bright sunlight streaming from a cloudless sky. The lava flows looked quite incongruous in this lush landscape of mountain meadows, aspen groves and pine trees. I tried to imagine what it would have been like when those hot and fuming lithic invaders had trespassed upon this pristine land.

As my gaze scanned the grassy meadow near at hand I noticed the many yellow wild flowers that were sprinkled through the long grass, waving in unison under the fresh breeze, as though in concert with the quaking leaves of the nearby aspens[11]. Below me I could see my car parked next to Louder Creek (really!) that meandered its way across the meadow, clearly in no hurry to spill over the escarpment

---

[11] Populus tremuloides

and exhaust into the desert below.

Perhaps it was due to the high altitude and thin, cool air but, as I stood on Brian Head, my usually well controlled emotions flooded to the surface and a shiver ran through my body. The beauty, serenity and solitude of the landscape surrounding me spoke directly to my heart, moving me deeply. I happily acknowledged how lucky I was to be there.

"What a magnificent place to go geologising," I said quietly to myself, unable to contain the euphoria I felt.

Continuing my reverie, with senses heightened in the bright sunshine, I heard the breeze shushing through the pine trees and thought about who else might have stood atop Brian Head in days gone by, listening to that same sound. Were they early Mormon settlers fleeing religious persecution in the East? Or did a Piute Indian chief stand there, arms folded, headdress feathers fluttering in the breeze, nodding imperiously as he surveyed his tribal territory? Whoever it was, would they have been moved as much as I was? Would they have felt the power of place that was sending shivers up my spine? Probably. No person with a beating heart could stand impassively in that spot. Returning to the present, I took a deep breath and exhaled slowly. The heady scent of pine drifted on the breeze.

Suddenly, in a loud voice, for there was no one to hear, my heart spoke out:

"It's good to be alive."

Reluctantly, after who knows how long, for time and space had merged into a continuum atop the hill, I retraced my steps back down the gentle grassy slope of Brian Head to the car. Five minutes later I had parked beside the nearest of the lava flows that I had seen from on high. Up close the lava seemed less sinister; just a jumbled pile of black rocks. The flows were still largely unvegetated, but here and there, where finer rubble had been washed into crevices, a fern could be seen unfolding, or a frail purple wild flower braved the chill wind. As I collected my rock samples systematically and recorded details in my yellow-covered field note book I heard a vehicle approaching and turned to see a pale green pick-up truck heading in my direction, with the insignia of the National Forest Service very evident on its doors.

My first thought was "Shit, I'm in trouble here." But then I remembered that Ian Carmichael had arranged clearance for me with the Forest Service and I was fully authorised to be there and to be doing what I was doing. Still, I was a little unnerved as the truck stopped and a uniformed ranger got out and walked over to me.

"Hi there, watcha doin'?"

I quickly explained where I was from and what my purpose was in being there.

"OK, so you're the guy. I did hear somethin' about a student doing field work up here.

Name's Sharon by the way."

He extended his hand as I identified myself, trying not to look nonplussed as I shook hands with a man called Sharon. But then I remembered that the Mormons of Utah are quite relaxed about restricting popular names to a particular gender. Sharon seemed like a friendly chap and we were soon chatting amicably. My Aussie accent singled me out as an exotic import and he showed genuine interest in what I was doing on his patch.

After answering his most urgent questions I queried him in return:

"Are there many fish in the streams up here on the plateau?"

"Yeah, there are rainbows and browns and maybe a few brook

trout but you won't find anything of size just here. If you want a decent fish you need to go down a bit, to where there are deeper pools."

"Be nice to give it a try sometime while I'm here," I said, in all innocence, "but I didn't bring anything to fish with from home so I'll have to see if I can get some gear down in Cedar City."

"Nah, you don't need to do that, I'll lend you a couple of my fishing poles. See what I've got here with me?"

We walked around to the rear of his truck, where he lifted a tarpaulin to reveal two already rigged fishing rods and a tackle box.

"Would you like me to show you the best fishing spot for miles?"

"Well that would be wonderful, but I don't want to impose …."

"No trouble at all. I'm just about to finish my shift so I can show you where it is on my way back to Cedar City.

Jump into your car and follow me."

Sharon led me along a rough dirt track that wound down a steep valley where one of the larger creeks (not Louder Creek) had cut into the edge of the escarpment.

"You see that pool down there with the big pine tree next to it?"

I nodded in confirmation.

"That's where you wanna fish. Take these poles and lures and give it a go. Just drop the gear back to my place in Cedar City sometime in the next few days. It's on Cedar Drive, three up from the corner of 7th Street. A white house with a driveway on the left hand side."

"Leave the stuff out back if there's no one home."

"Thanks very much. It's really kind of you. But I really only need one of these."

"Nah, take two, it'll save you changing the rig if you get hooked up on something."

Thanking him again for his generosity and trust, I bade him farewell and set off down to the pool to spin for trout with his fancy rods. I think you could say I was in seventh heaven!

***

Sharon's Pool (as I thought of it) did not disappoint; fishing until it was almost dark, I caught three wriggling rainbow trout and two bobbing browns. Elated, I returned to the car and drove back to the National Forest campground where I had set up base camp for the duration of my visit to the area. Fresh grilled trout was definitely on the menu that night.

Late the next day, after diligently mapping and sampling another lava flow, I revisited Sharon's Pool and had another successful fishing episode, all the while in total solitude.

On the third day, I needed to go into Cedar City for more provisions and grudgingly accepted that while there I should return Sharon's fishing gear as he had instructed.

I found Cedar Drive and its intersection with 7th Street and looked for the house as he had described it. There was some confusion about which house was his – the layout did not seem to fit exactly the simple description he had given me a few days before – but eventually I settled on the one I thought must be it and knocked firmly on the door. There was no answer; apparently, no one was home. So, as instructed, I placed the fishing rods and tackle box at the back of the house, where they could not be seen from the street and left a note of thanks.

***

Some weeks later, back at Berkeley, a letter arrived at the Earth Sciences Department, addressed to the …

"Student doing geological field work in Utah".

The ladies in the office knew that I was the only one fitting that description at the time, so the letter was handed to me.

To my utter dismay, Sharon had not recovered his fishing gear.

Again and again I thought through the instructions Sharon had given me. Had I left it at the wrong house? No, it must have been the right house; there were just not that many possibilities. Had someone seen me go in there and later stolen it? Perhaps. The fact was, the fishing poles were gone; at least, that is what the letter said. Anger, tinged with guilt, pulsed through my mind as I pondered what to do. I knew I should recompense Sharon for his loss, even if it was due in part to his very casual and imprecise description of where to leave the blasted rods. If I had indeed followed his instructions correctly, as I believed I had, was it my fault if some bastard came in later and took his gear away? The whole affair was very distasteful and unfortunate; true enough, but in my mind, I began to rationalise the matter, as the days slipped by.

***

Now, fifty years later, I still look back on that episode with genuine regret. I don't know why I failed to respond to Sharon and why especially I did not at least offer to cover the cost of the missing fishing gear. Perhaps in the bustle of busy Berkeley the Utah experience seemed unreal, dreamlike, detached from reality. Whatever the reason, and there can be no excuse, I confess that I betrayed the trust and kindness of a generous stranger. I am truly sorry Sharon.

Ever since then I have striven to be diligent with borrowed articles, to return them faithfully in good time. But the simple fact remains: A cherished memory of nature at its most sublime carries a black stain that time will not wash away.

8
REFLECTIONS

AETHER

We sat in the rooftop bar of the Hilton Hotel in London, sipping champagne and feeling very pleased with ourselves. Well, I was the one elated by self-congratulation and Margaret was kind enough to indulge my pride. It was late September, 1970, and we had that day received word that my PhD thesis had been accepted. My degree would soon be granted and I would then officially be Dr Lowder.

The news put a seal on the momentous, life-changing experience of post-graduate study at the University of California, Berkeley. After breathing the heady atmosphere of academia for four years, I was about to become the irrepressible Ian Carmichael's third successful PhD candidate. Many others were to follow later, but as one of the first, I knew I held a special place in the Prof's heart. I was confident that what had been learnt under his tutelage would serve me well long after the sights and sounds of one of America's greatest institutions had faded from memory.

But nothing had faded by that September night in London as we celebrated my new status. After the visit to Europe, and the forthcoming Christmas at home in Sydney, we were bound for Adelaide. Before leaving Berkeley, I had procured a job as a petrologist at the well regarded contract minerals research and service organisation called Australian Mineral Development Laboratories, better known as 'AMDEL', located in Adelaide. I was due to start in January 1971 and AMDEL was funding our relocation from California to South Australia.

In some ways, our celebration at the Hilton was a bittersweet moment. My efforts and Margaret's hard work at Blue Cross had reached a successful conclusion in Berkeley. That was true enough, but now we had to move on from that detached, insulated ivory tower, where I had become used to imbibing the aura of excellence emanating from my peers. Now we had to start adjusting to the realities of earning a living and settling into some sort of normal urban life. In Berkeley, we had often felt special, different, even exotic, having been treated as such by the Sales and other friendly Americans:

"Oh, that is so cute! Say that again." It was as though we were domestic pets; talking budgies.

In Berkeley we had been curiosities, our foreign status singling us out for special attention. In Adelaide we would be invisible, just one of the crowd, with accent and national origin no longer affording us celebrity or even curiosity status. In Berkeley, my ego had been nurtured, even if sometimes abusively, by Ian Carmichael. Who would massage my self-esteem in Adelaide? In Berkeley, I had luxuriated in a milieu of privilege and intellectual stimulation. What could Adelaide, a place I had never been to but by reputation pretty dull, offer as a substitute? It was therefore with some trepidation that we talked about what might lie ahead of us when we finally got to Adelaide.

"We don't know the first thing about Adelaide really," Margaret observed, as she gazed down at Buckingham Palace, its lights shining brightly through a clear London evening. The vibrant city that surrounded us seemed to mock our mental images of the South Australian capital; the antithesis of what we thought we would encounter in the antipodes[12].

---

[12] Ah, the crassness of youth! Adelaide turned out to be wonderful.

"No, we don't, but it can't be all that bad. Amdel looks like an interesting place to work and the salary is pretty good[13]. We should be able to buy a home and become Mr and Mrs Ordinary Australia. Or rather, Dr and Mrs.

Maybe we'll even hear the patter of little feet in due course."

"I'm ready!" replied Margaret.

"I'll miss Berkeley though," she mused.

Breaking through the melancholia, I reminded her that we had just spent a wonderful, and surprisingly sunny, month touring Britain and were soon to cross the Channel to travel around continental Europe for another two months. In those days, Arthur Frommer's *"Europe on $5 a Day"* was the budget travellers' bible. And we used it religiously! For the most part, we managed to acquire food and accommodation for two within that budget ($5 a day each). Drinking champagne at the London Hilton was definitely not part of the $5 a day regimen, but then, it was not every day that a Mister became a Doctor.

"I'll miss Berkeley too, but I'm ready for something new."

"What's really going to seem very strange once we get to Adelaide though is our time in Papua New Guinea, traipsing through the jungles and living in *haus kiaps*; not to mention the oddball characters we met."

Our base had been in Talasea, a small government outpost on the island of New Britain, north of the Papua New Guinea mainland. Part of our time had also been spent in two of the remote native villages up the coast, near the large caldera that was the prime focus of my field research.

"Here's to Bob Heming in Rabaul," I toasted, "to Gari and Utu in Bulu Muri, to Gabriel in Voganakai, to Sung-in at Volupai – bless his

---

[13] My starting salary was $8,300 p.a.

thieving fingers – and here's to Wilhelm and Eddy and the Humphreys and all the other peculiar people in Talasea."

"They may have been a bit odd," responded Margaret, "but without their help our task would have been impossible."

Over another glass of champagne ("What the hell! It's a special occasion!") we began to reminisce about those remarkable three months in 1967, immediately after we were married and before Margaret joined me in Berkeley. Lubricated by the champagne, our conversation drifted nostalgically back to that time.

"Do you remember the shock we felt on arrival in Rabaul?" I asked Margaret.

"I sure do. Especially that old British soldier sitting and drinking gin day after day in Leo's Guest House? What a character!"

Agreeing with Margaret, I added, "we'll never be able to relive those days but thanks to the diary you kept, we'll never forget them either."

"Here's to you, Dr Lowder, or should I say *Masta Bilong Brukim Ston?*"

2. "MASTA BILONG BRUKIM STON"

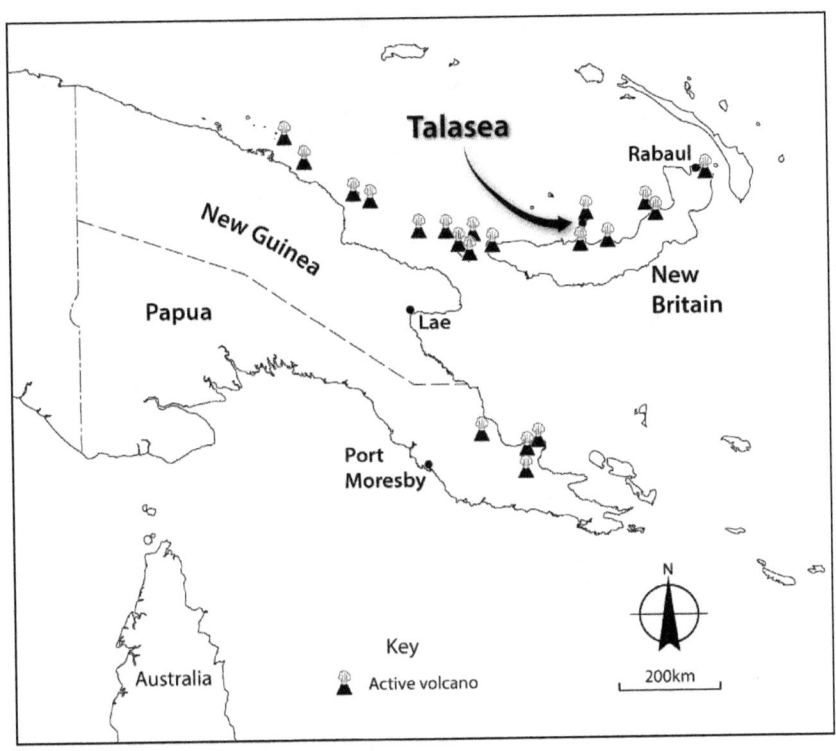

9

RABAUL

EARTH

Papua New Guinea (PNG) is the leading edge of the Australian continental plate, which is steadily making its way northwards at the rate of about 6cm per year, or about as fast as your fingernails grow. As it does so, it interacts aggressively with the Asian plate, and other smaller plates to our north, giving rise to the volcanic and earthquake activity that characterises the Pacific Ring of Fire, of which Papua New Guinea is part. The high mountains forming the backbone of PNG, and its western neighbour, Indonesian Papua, are in effect the Australasian equivalent of the Andes in South America. The highest point on the Australian continent (Oceania) is not Mt Kosciusko in New South Wales, a mere 2,228m (7,310ft) above sea level, but Mt Carstenz (*Puncak Jaya* in Indonesian) at 4,884m (16,024ft), in Indonesian Papua. To the east, in Papua New Guinea, Mt Wilhelm is the high point of the range, rising to 4,509m (14,793ft).

It was to this active plate margin that I had been directed by the pioneering New Guinea geologist, Norm Fisher. Dr Fisher was Director of the Bureau of Mineral Resources[14] in Canberra when I wrote from Berkeley in early 1967 seeking his advice about an area for field study of volcanoes in PNG. Not that

---

[14] Its full name was Bureau of Mineral Resources, Geology and Geophysics, but this was usually shortened to just BMR. The equivalent body today is Geoscience Australia.

crustal plates, subduction zones and sea floor spreading were part of the scientific vernacular at that time. The Ring of Fire was a well recognised but poorly understood phenomenon that I was to play a small role in elucidating as the theory of plate tectonics evolved in the late 1960s.

My interest in the volcanoes of PNG had first been sparked by tales my brother, Graham, had told me of his adventures flying light aircraft in PNG in the 1950s and early 1960s. At that time it had not occurred to me that I might one day be there myself, getting up close and personal with some of those fiery mountains. When, as a PhD candidate at Berkeley, the opportunity to do so began to take shape I had no idea where to start. I was happy therefore to accept Dr Fisher's recommendation that I direct my field work to the Willaumez Peninsula in New Britain. The arc-shaped island of New Britain lies to the north of the PNG mainland and the area he was recommending juts out from the north coast of the island as a peninsula, 60km long, 20km wide at its base and as little as 4km wide farther north. The small government outpost of Talasea[15] was situated about half way up the east coast of the peninsula and became our operational base for the three months we spent there.

It was known that the Willaumez Peninsula consists of a chain of overlapping, geologically young volcanoes, trending northwards from the rest of New Britain, and culminating in a large caldera at its northern end. Inside the caldera there is a fresh water lake with a resurgent volcano, which last erupted in 1895, emerging from the water like a rising Phoenix. Other than that, virtually nothing was known about the geology of the area – it offered me a totally clean geological slate on which to write my academic future.

---

[15] Pronounced Tal-a-see-a.

New Britain is close to 500km long and the main settlement in New Britain when I visited, really its only town, was Rabaul, at the northern tip of the island. Rabaul was already well known as an active volcanic zone and the Australian Government had established a volcanological observatory on the rim of the caldera that surrounds Rabaul and forms the deep and attractive Simpson Harbour. From the observatory, there was an uninterrupted view across the town and its harbour to Tavurvur volcano, the principal source of modern activity at Rabaul, with a history of regular eruption. In 1994, a large eruption at Tavurvur and its neighbour across the harbour, Vulcan, ejected vast volumes of ash that rained down on Rabaul and effectively destroyed the town. It has not been rebuilt, although some enterprising people moved back after the eruption and staked their claim to ground while the ash was still warm.

***

Fortunately for Margaret and me, all that lay in the future when we stepped out of the Fokker Friendship into the heat and humidity of Rabaul airport, which was tucked away just under Tavurvur, on Friday, 21 July, 1967. The scene was one of barely controlled chaos as the plane was unloaded and we retrieved our baggage. There to meet us was Bob Heming, Acting Volcanologist-in-Charge at the Observatory, who took us to Leo's Guest House, the only respectable cheap accommodation in town. Leo's was pretty basic but, at $3 per night each for bed and breakfast, it suited our very limited budget; and the entertainment was free. Chief performer at Leo's was a British expatriate, once a soldier in India but by then a recluse in Rabaul, who sat in the dining room every day, shoeless, his thin bare legs crossed, drinking gin and lamenting the decline of the Empire and the indolence of the natives. From Leo's it was a short walk to the "Bung" – the native market where fresh fruit and vegetables and local handcrafts were sold. Most of the vegetables were assembled into lots for sale at a shilling (10c) each. Margaret was especially fascinated with the *bilum* bags on sale everywhere. They were a kind of hand woven string bag used by the women for carrying personal items and Margaret bought a couple of them to hold the products of what was essentially the first grocery shopping of her married life.

"I'm not going to carry them like the *meris* do though," she insisted, watching the native women drape the *bilum* bags over their heads, with the straps around their foreheads and the laden bags hanging behind their backs as they leaned forward under their heavy loads.

"Why not?" I asked, grinning, "you're young and strong, and at least your boobs won't dangle as much as theirs do."

An unladylike grunt was all the reply I got.

Discretion being the better part of valour, I elected not to pursue this line of discussion any further as I guided my young wife carefully through the maze of goods laid out on mats of palm fronds and left the Bung. Next stop was a Chinese trade store where we bought cheap cotton clothing for field work. At a pharmacy, we topped up our medical kit with penicillin (available without prescription) and other medicines that were to prove very useful before long.

During World War II Rabaul was a major base for the Japanese occupation force, which invaded in January 1942 and decimated the small contingent of Australian defenders. Over 1,000 Australian soldiers were taken prisoner and 160 of them were later massacred by bayoneting and shooting. At least 800 soldiers and 200 civilian prisoners, most of them Australian, died when the ship on which they were being transported to a prison camp on Hainan Island, the *Montevideo Maru*, was sunk off the coast of Luzon in the Philippines by a torpedo from the American submarine, USS *Sturgeon*. The Japanese plan was to make Rabaul, with its superb harbour, the launching pad for the invasion of Australia.

The occupying forces used Rabaul's geology to good effect, forcing 600 British soldiers captured in the fall of Singapore to dig numerous tunnels and caverns into the soft volcanic ash deposits. Only 18 of those soldiers survived. It has been estimated that the underground openings amounted to around 450km in total length. Regular bombing by the Allies from 1943 to 1945 effectively obliterated all the structures erected in Rabaul but the presence of so many tunnels made the town virtually impregnable regardless. They were the key reason that Rabaul was never captured by the Allies in battle. Late in the war the Americans even took to bombing the dormant volcanoes around the harbour in a futile attempt to make them erupt. When Japan surrendered in 1945 it was reported that there were still 69,000 Japanese troops in Rabaul. As Bob Heming drove us up the road to the Observatory, just 22 years later, we stopped to look at some of the tunnels dug into the soft ash of the caldera. One such tunnel contained (and still contains) five landing barges, end to end, and in another, British Admiralty maps overprinted in Japanese had been found.

"How the hell did they get those barges in there?" My query was an echo of every other itinerant visitor to the site over the preceding 20 years.

The bulk of our field gear had been packed into a trunk and sent to Rabaul from Sydney aboard the SS *Aros*, which was due to arrive three days after we did. This of course did not happen and we finally were able to retrieve our shipment only on the following Friday, a week after our arrival. Stuck in Rabaul longer than expected, we made good use of the time to prepare for Talasea. This included setting up an account at "Steamships", a rival to the legendary Burns Philp & Co store, so that food and other supplies could be ordered (by ordinary mail – there was no Internet back then!) and sent to us in Talasea.

Bob Heming took one look at the field gear we had brought from Sydney and declared:

"That'll never do. You'll die trying to sleep on Li-Los[16] in a tent with a floor like that. Up here we use just a large tent fly and have the boys make stretchers with canvas loops over local timber poles. I'd better set you up properly."

With that, he produced a couple of metal patrol boxes (with long looped handles through which a carrying pole could be inserted), a big fly, a couple of stretcher canvasses and, just to be on the safe side, a shotgun and cartridges, which would complement to .22 rifle I had brought from Sydney.

"Come out the back and we'll try this thing out," said Bob, picking up the shotgun and leading me out through the rear door of the observatory and down about twenty metres below the ridge line, where we were shielded from Rabaul town. In a clearing that looked across a small wooded valley, below the outer edge of the caldera, he handed me the shotgun with the assurance:

"You can fire this from here; shouldn't bother anyone."

Fat chance! As I would soon learn, in New Britain there were very few places that you could do anything undetected, least of all fire a gun. Carefully, I loaded a shell, closed the breech and, aiming high in the air, squeezed the trigger.

Bang! I got quite a shock, but not as much as the native people in the bush just across the valley! As the much louder than expected report echoed round the valley a dozen or more natives emerged from the undergrowth, yelling and gesticulating wildly. Bob and I tried to

---

[16] Air mattresses.

look innocent as we skulked quickly back into the observatory.

Preparations for departure from Rabaul continued and included leaving most of our inappropriate gear brought from Sydney with Bob at the Observatory, where the air conditioning would protect it from mould. We repacked our trunk and one of the borrowed patrol boxes and arranged for them to be shipped to Talasea on board the coastal vessel M.V. *Motoko*. The other patrol box and the rest of our gear would fly with us to Talasea on Saturday.

Margaret took naturally to the role of logistics organiser and my initial doubts about bringing my young bride to such a place soon subsided. She also proved to be an excellent chronicler, keeping a diary of each day's events throughout the time we spent together in Papua New Guinea. When, decades later, I transcribed her notes and observations, written by hand in a school exercise book, I could not help but shake my head in wonder and declare, time and again:

"I can't believe we actually did that!"

But, of course, we did!

Margaret goes shopping at the 'Bung' in Rabaul

AIR

Happy Birthday

Saturday, the 29th of July 1967, was my 23rd birthday. Margaret had given me a bottle of Scotch and …

"I'm giving you Talasea as well, once we finally get away from this soggy place."

The TAA DC-3 stood on the tarmac at Rabaul airport, looking forlorn in the pouring rain. I watched as freight was loaded into the interior. Where were they putting it all? Soon I knew the answer as we and the other travellers were called to board: Passengers sat in the right hand seats; freight was stacked on and between the seats all down the left hand side. The seats were canvas; not comfortable but then,

"This is not going to be a very long flight, is it?" Margaret asked hopefully.

Airborne and immediately we were into the clouds. Climbing, climbing, climbing, we headed south, towards our first stop, Jacquinot Bay, on the southeast coast of the island. Above the rain clouds but under higher overcast the ride was reasonably smooth. I couldn't help wondering … how is the pilot going to find the airport at our destination? There was no air traffic control to speak of after leaving Rabaul and I very much doubted that landing aids were installed at the remote airstrips. I just had to trust to the training and instinct for self-preservation that are mandatory requirements for aviators in Papua New Guinea. Most of the pilots were young men from Australia, gaining experience and building up their hours before seeking a job back home with Ansett or TAA. Ten years beforehand, my older brother, Graham, had been one of them.

After more than an hour we should have been nearing Jacquinot Bay but were still stuck above one layer of cloud and below another. As the aircraft turned, circled and turned again I thought: Is the pilot lost? I could have sworn we were headed back towards Rabaul. Then, as we circled yet again, I understood the pilot's strategy. Slowly, as the aircraft spiralled downwards through a gap in the clouds, whitecaps on the grey sea became visible beneath us. We headed south again, just below the cloud base. At this low level, the ride was far from smooth and it sure was bucketing down. Rain water streaked in rivulets along the windows as we skimmed over the waves at barely three hundred feet. Another half hour of jolting and shaking and I could see the rectilinear patterns of coconut plantations as we approached Jacquinot Bay. Finally, we landed and taxied to the terminal – a makeshift tin hut under which several Europeans huddled, sheltering from the downpour. The natives stood outside, looking miserable and shivering in the rain. I looked at Margaret beside me; her grim white face told me this is no joy flight.

"Is it always bloody raining in this place?" we heard the pilot ask, as he supervised the unloading of freight bound for Jacquinot Bay.

"Only during the southeast season," answered a local expat, "but we do get 260 inches a year[17]."

"Bloody hell! How do you survive?"

After this exchange Margaret asked me,

"Do you think the pilot's been here before?"

"I'm sure he has. I guess he's just as incredulous as we are that it can rain so much and yet people still manage to live here."

The southeast season is when the prevailing wind comes unimpeded

---
[17] Or about 6.5 metres of rain.

across the Solomon Sea from the southeast and dumps its moisture on the south coast of New Britain. Shielded by the mountains of the interior, the north coast, including Talasea, is relatively dry at this time. There, I was told, they cop the rain in the shorter northwest season, from November to February. Good reason for us to be finished and out of Talasea as planned by the end of October.

We were airborne again, minus three passengers and some freight. Circling over the sea we spiralled upwards through the clouds, even higher than when we left Rabaul. We had a mountain range to cross and in that weather, we needed to give it good clearance. The rhythmic drumming of the DC-3's two reciprocating engines was reassuring, almost soporific, with not even a hint of missing a beat. I continued to listen to them very carefully but began to understand why this aircraft was so highly valued during the war[18]. Graham, my pilot brother, a veteran of aviation in PNG, which included flying the DC-3, had told me that they are the most airworthy and reliable aircraft I am ever likely to fly in. As we sat, suspended in the aether, I was happy to believe him.

Suddenly the cabin was filled with light as we broke out of the clouds and turned west, across the island, towards our next stop, Cape Hoskins, on the north coast. Here and there I could see green-clad mountain peaks poking through the clouds, as though popping their heads up to see what is disturbing the peace. Landing at Hoskins presented no dramas as the sky was pretty much clear and once on the ground the unloading ritual was repeated: suitcases, mail bags, boxed orders of bread, larger packages of frozen food, and several egg cartons, tied up in twos. Just before the joining passengers boarded, one of the flight attendants walked down the aisle carrying a single egg.

---

[18] The military version of the DC-3 was known as the C-47 by American forces and the Dakota by British forces.

"You can't say they're not thorough!" declared Margaret, shaking her head at the spectacle.

Then it was the final leg, a short hop across Stetin Bay to Talasea, where the airstrip ran east-west across the peninsula and the terminal was a copy of that at Jacquinot Bay. Here, though, it was sunny as I looked out at the small group of expats meeting the plane. Among them was Wilhelm Speldewinde, the *Kiap*, or District Officer, who was expecting us and greeted us warmly as we stepped off the plane.

"Welcome to Talasea," he said, shaking our hands before ordering some of the natives, in Pidgin, to carry the freight and baggage into the shed that served as a terminal.

"How will I ever learn to speak that language?" I asked myself quietly, as I watched.

"I'm just glad we're here at last," declared Margaret, whose relief to be on solid ground was palpable.

While I waited for our baggage to be unloaded, staying close to the aircraft to make sure nothing was missed, I looked around me. Near at hand was a tennis court, creepers growing profusely through its wire fence. Beyond that lay Volupai Plantation, with its countless coconut trees lined up in neat rows. Looming over all was the unmistakable shape of a volcano. I had an uncomfortable sense of being watched, as if the mountain was looking down at me with amusement, or perhaps disdain, even pity, from under its cloak of deep green rain forest. It felt as if the volcano knew what was in store for me, while I stood there, a complete neophyte in that environment.

Wilhelm broke into my sombre mood:

"I'll take you into town now," he said, "don't worry about your stuff, the boys will bring that in with the tractor and trailer."

"You are about to discover the delights of the Talasea Club," he announced, the irony in his voice only adding to my anxiety.

Delights they may be but at that moment all I could think about, as I looked up at the forest and the volcano, was: What challenges lie ahead?

"What the hell was I thinking?" I whispered to Margaret, as we were driven into Talasea. I felt very apprehensive indeed about my choice of field area and the difficulties I would face carrying out geological work in this primitive, jungle-clad environment. Up till this time it had been a dream, an aspiration, a figment of my imagination. Now there was no escaping the harsh reality that confronted me, confronted both of us. We were here. The time was now. Together we had to survive the next three months while I explored this peninsula, endured its climate, subdued its rain forest, conquered its mountains, dealt with its people and unveiled the hidden treasures of its geology.

As Wilhelm's Landrover entered the small settlement of Talasea I got my first whiff of "rotten egg gas" – the putrid but curiously sweet-smelling gas that leaks from the ground in most volcanic fumaroles. What kind of people would choose to live in such a strange and malodorous place?

WWII Lockheed Hudson bomber on Talasea's pre-war airstrip

10

THE TALASEA CLUB

"You look like a bloody clown!"

Harry's less than flattering description of his wife, Thelma, rose above the hubbub from the other guests, who studiously ignored the insult, as we entered the Lounge of the Talasea Club. Margaret and I looked at each other and rolled our eyes; what sort of place had we come to? Yes, Thelma did overdo the lipstick and rouge a bit, well, quite a bit really, and her ample figure was dressed rather garishly, but then, this was supposed to be a fancy dress party so what did Harry expect?

We had been told that the Humphreys were the leading lights in the tiny Talasea community. Harry managed the district's largest commercial enterprise, Volupai Plantation, situated just west of Talasea, where we had landed at Talasea's airfield only hours beforehand. Originally English, he dominated the local community as much as any of his English ancestors might have done their stately manors. Thelma, meanwhile, set the social agenda for everyone else and nothing of significance could happen without Thelma's imprimatur.

The Talasea Club was the social gathering place for the local community, much as a local pub or RSL club might have been in other, less isolated regions. The "Lounge" was a small square building, up on stilts, of lightweight construction, with louvre windows that looked down onto the harbour. The community owned it and had also built four accommodation rooms in a separate wing to house visiting government officials and other workers, who would otherwise have been billeted with residents. The four rooms each had two single beds and opened off a veranda, at the end of which was a small kitchen and a bathroom. Meals were taken on the veranda. A screened opening high up at the back of each room kept the insects out but allowed the rain to blow in. Pretty basic accommodation, certainly, but then, this was not exactly downtown Manhattan. We were occupying one of the rooms on an interim basis but at $6 a day for two for full board, it was beyond our budget for the longer term. We had already asked whether we could camp in the grounds for a nominal fee and use the bathroom while fending for ourselves in the kitchen at mealtimes. An answer to this request was pending.

"Perhaps we could have planned this part of the expedition a bit more carefully," I had admitted to Margaret earlier, as we stood on the veranda, gazing at the vista before us.

"I thought you took care of all that before you left Berkeley," she replied, more than a little incredulous that I had left so much to chance.

It was not that I had deliberately left it to chance. In fact, I had not thought much beforehand about how we would survive once we actually got to Talasea. I mean, we did bring a tent from Sydney after all, albeit one that turned out to be quite unusable in the tropical environment. Although I had been living in America for ten months, the Aussie maxim of "She'll be right!" must have pervaded my subconscious.

The Talasea Club occupied an idyllic position. Situated on top of a hill, it overlooked the administrative offices, wharf and Garua Harbour, including the pretty, jungle-clad island that partly closed the bay and made it into a harbour. The picture post card view was framed by coconut palms that waved in the fresh breeze blowing in from the Bismarck Sea. It was the sort of place that, in more settled climes, would be coveted for an up-market resort full of fat, sunburned tourists complaining about the service. Coveted except for two things: sand flies and hydrogen sulphide.

For some reason Talasea (alone in the district) was infested with an annoying species of sand fly that had a nasty bite but which, fortunately, was a very low flying variety – it never bit its victims above the knees and was only really a problem at night time. The local expats had devised an effective defence against the little blighters, or biters – Johnsons Baby Oil, smeared from ankle to knee was all that was required to prevent an attack by the sand flies. This was explained to us by Wilhelm, as he introduced us to the other members of the Talasea community.

"I wondered why they all had shiny legs," Margaret observed in a quiet aside.

The other stain on the character of this beautiful place was less painful but more inescapable. The District Office had been established at Talasea because it had an excellent small harbour. In putting it there, however, the PNG Administration had paid little heed to the active geothermal field that permeated the area, with boiling pools, bubbling mud pots and the pervasive aroma of rotten eggs. Hydrogen sulphide, or "rotten egg gas", escaped from every orifice in the ground and then invaded every orifice offered by the unfortunate humans who were forced to live and work there, all of them expatriates, for the native population was too smart to tolerate such indignity.

Shyly, we joined the Saturday night party that was the highlight of the week for the residents who crammed into the small lounge area of the club. Fancy dress was indeed the order of the day and most of the locals had spared no effort in dressing for the occasion. It also partly explained Harry's solecism as the Humphreys had arrived at the venue; only partly though, for, as we would learn later, Harry was prone to strong opinions and was not afraid to express them. The preparations for our sojourn in Talasea had not anticipated a need for fancy dress, quite reasonably I thought. The best we could do was to construct cone-shaped paper hats with red marks dribbling down the side – a feeble attempt to portray the volcanic objective of our presence.

"I feel bloody ridiculous in this thing."

"Me too," Margaret agreed. Soon after both "volcano" hats were discreetly placed on the floor and pushed back out of sight.

As the party continued we struck up conversations with key players in the social hierarchy of Talasea. We met Dr Stewart Bartle and his wife, a very friendly couple, about fifteen years older than us. We also met the parish priest from Bitikara Mission, located between "town" and the airstrip at Volupai. And we met Eddy, the PNG-born Chinese man who ran the trade store in Talasea; Eddy and his wife were afforded honorary "expat" status but were always on the fringe of local society. As the evening wore on, we got the feeling that we were being vetted by the Talasea community. At 10.00 pm we left the party, quite exhausted after a long and eventful day and collapsed into our beds in the accommodation wing. Later we heard that the party had continued until 1.45 am, but we were so knackered that it would have required a volcanic eruption to disturb us.

The next day we were invited to play tennis at Volupai, clearly a mark of social acceptance. While there, Harry Humphreys summoned us over.

"We can't let you camp at the club," he announced, with obvious authority that drained my hope, but then he went on:

"But if you like, you can come and base yourselves here at Volupai. We've got a house that's going to be vacant from Tuesday while one of our staff goes on furlough. You can stay there while you're here. There's also a house-boy who'll look after you."

"That sounds perfect," I responded, "but I just wonder whether we can afford the rent for three months."

"It won't cost you anything, really," Harry replied, with some irritation, "you'll just have to order your own supplies from Rabaul and fend for yourselves. And you'll need to tip Sung-in a few dollars each month as well."

We were overwhelmed. Such a generous offer was way beyond our expectations. I think he was rather pleased with himself for being able to make this offer and it suited his image to be seen hosting us. We may not have had the social standing of a District Commissioner or some other high ranking government official, but we were there in the cause of science. Harry was not about to share with lesser mortals any kudos that might emerge from that. Clearly, we were the most exotic and unusual visitors this remote backwater had seen in a long time.

"Thank you, you're very kind and we'll gladly accept your offer. We promise to take good care of the house while we are there."

"It's agreed then. I'll send someone in to pick you up on Tuesday."

I watched as Harry's rotund figure made its way back to the tennis shed, still shaking my head in disbelief at our good fortune. The offer of a house, complete with house-boy, at Volupai solved what was looming as a major issue for us at Talasea. Looking back on it, I am astounded at our naivety in turning up there without arranging something like this in advance. Oh, the confidence of youth!

***

With our accommodation needs solved we spent Monday exploring the area around the settlement. It was a warm, relatively humid day, like most during our stay, but quite tolerable so long as the breeze was blowing. New Britain had seen much action during World War II and Talasea had been right in the thick of it. First stop was the old, pre-war airstrip, up on the hill just behind the club, where the remains of several aircraft sat decaying in the tropical climate. They appeared to have used the strip for emergency landings, as it was much too short for take-off. At one end lay an American Hudson bomber and near it a heavily damaged Lancaster bomber, while at the other end of the strip, almost lost in the bush, was a Japanese Zero. God only knows what happened to the unfortunate crews of these aircraft. Another Zero wreck sat on the reef, just off the point, alternately appearing and disappearing as the tide ebbed and flooded.

Later in the afternoon we strolled the kilometre or so that was required to visit the priest we had met on Saturday night. Bitikara Mission consisted of an old weatherboard house that was quite substantial, surrounded by a cluster of smaller outbuildings, one of which constituted a school. We understood that the mission's main house had been used by Japanese occupying forces during the conflict and was the scene of very active fighting.

"See, here are some of the bullet holes from those days," said the priest, proudly poking his finger into several ragged holes in the wall.

Watching and listening to the priest really brought home to us the reality of the fighting that took place right there just 22 years beforehand. As we followed the priest out onto the veranda for a cool drink we found it really hard to imagine that so much blood could have been shed in such an idyllic setting.

Talasea was not the only place where the ugliness of war had desecrated such natural beauty but, unlike the temperate climes of Europe, here the verdant jungle had already covered up the scars.

On the way back from Bitikara we dropped into the small hospital building ("*Haus Sik*" in Pidgin) to say hello again to Stewart and his wife. Then it was back into Talasea proper, inspecting the wharf and taking a first professional look at some of the geothermal features scattered through the vicinity. None of these was very substantial or violently active, fortunately, as otherwise Talasea would have been uninhabitable. But there were many gently bubbling hot springs and quietly fuming vents, some just a crack in the ground or a small black hole, no bigger than a rabbit hole. Down on the shoreline hot water dribbled out of the ground at low tide, depositing traces of yellow sulphur as it flowed over the rocks and sand into the harbour. And everywhere the subtle stench of warm rotten egg gas, which, though highly poisonous, is easily detectable by the human nose at concentrations well below the toxic level. Offensive as it is, the gas imparts an almost sickly sweet sensation. To this day, like a Wagnerian *leitmotif* evoking Siegfried, the smell of eggs brings to mind cherished memories of Talasea.

One such memory is of *mimosa pudica*, an invasive weed (a legume) that seemed to thrive in that smelly environment. Originally from the Americas, this groundcover plant is sensitive to touch – as you brush it, the leaves close up and the stems droop for several minutes; I liked to think of it as "playing dead". It was the first of many encounters with strange and exotic plants that were to take place at Talasea.

***

Tuesday saw us move into the house at Volupai, where Sung-in, the *haus-boi*, helped us unpack and then set about preparing lunch.

Sung-in did not speak English but seemed able to unscramble our garbled mixture of English and Pidgin and anticipate our needs. He turned out to be quite a good cook of simple fare. Margaret decided he needed to learn to use herbs and spices to add a bit more flavour to his offerings. And I decided it was high time to focus on learning enough Pidgin ("*tok pisin*") to be able to communicate.

The next day we set off to Volupai village, about an hour's walk south along the coast from the plantation of the same name. Along the way I started to look more closely at the rocks outcropping along the shore and was immediately puzzled by what I saw. With no prior geological mapping in the area, everything I saw was new and posed questions for my limited volcanological experience. Sung-in had arranged for us to meet Levi, who was looking for work as a guide and carrier. We found Levi, or rather, he found us – we tended to stand out in the crowd – and we knew it was Levi because he had his name tattooed on his forearm. Well, actually, he had the mirror image of 'LEVI' tattooed on his arm and Margaret and I spent some time trying to figure out how this had happened.

"He must have done it himself by looking in a mirror," Margaret deduced.

After arranging for Levi and Peter (his friend – "*pren bilong mi*") to turn up at Volupai plantation at 7.30 am on Thursday, we walked on past the village to examine the outcrops on the rather prominent headland we could see ahead of us. Trailing along behind us was a veritable tribe of children, all fascinated to see this "*Masta*" and his "*Missus*" walking, not driving, through their territory. Much chattering and excitement surrounded us and we felt like celebrities. Until, that is, we made haltering statements to the children in our rudimentary Pidgin, causing the kids to dissolve in laughter.

"Hmmm, a bit of work to do there," I admitted to Margaret.

At one point, I stopped and used my geological hammer to crack open some rock outcrops; the chattering ceased immediately and puzzled expressions spread across the little brown faces. This was a totally new experience for them and we could tell they were very curious as to why I would do such a thing.

"I speak English," announced one of the older boys, "what you doing Mister?"

After a few false starts, I managed to convey that I was a geologist here to study the volcanoes and their rocks in order to write a book[19] about the area. The children seemed quite satisfied with this explanation and I acquired there and then an epithet that was to follow me throughout the Willaumez Peninsula:

"*Masta Bilong Brukim Ston*" (Mister who breaks rocks).

The Talasea Club,
accommodation on the left, lounge on the right

---

[19] The concept of "thesis" was just too hard to explain – I used the simpler concept of a book about Talasea numerous times over the next three months.

11

VOLUPAI

In 1967 palm oil was just beginning to make an impact on the agriculture of Papua New Guinea. In those days, the plantations were still devoted principally to coconuts that were harvested and dried to produce copra. Cocoa was a secondary crop, grown in smaller trees between the endless rows of tall coconut palms. Volupai was the largest and most successful of the plantations established on the Willaumez Peninsula, a fact of which its manager, Harry Humphreys, was very proud. From their sprawling residence on Volupai Harry and Thelma Humphreys were the reference point for any new initiative that arose within the small community. Later, in 1975, when Australia granted independence to Papua New Guinea (Whitlam could not wait to get the PNG Territory off his hands) Harry was elected to the first PNG parliament in Port Moresby.

Harry Humphreys had a strong, even domineering personality, which was no doubt an asset in the remote and barely civilised place in which he lived. The plantation had a couple of expat workers, including Richard, whose house we were occupying, but other than that, the workforce comprised natives from Volupai and other nearby villages. Their tasks ranged from slashing the quickly growing grass with the machetes (or *"bus naips"* as they were known in Pidgin) that every man (*"boi"*) carried, to collecting, de-husking and splitting coconuts so they could be dried and the dried fruit (copra) extracted. Crude kilns were set up to aid the drying process, fired by coconut husks. The acrid smell of smoke from charred coconuts became a familiar, almost welcoming aroma as I walked back through the plantation towards our house after a day in the bush.

Our house on Volupai Plantation was quite a large, rambling affair, with open verandas and lift-up shutters rather than glass windows in most rooms. These were well suited to catching the breeze that drifted in from the sea each afternoon and cooled us down – there was no air conditioning! The kitchen was rudimentary but functional enough, with a large fridge and freezer for food storage. Electricity was provided by the plantation's generator and water came from rainwater tanks attached to the building. The sitting room offered several comfortable arm chairs and the bedroom had a lockable door. Probably the best attribute of the house was Sung-in, the *"haus boi"*, who cooked, cleaned and washed clothes reasonably efficiently. He was not a local, coming as I recall from Bougainville, but he seemed to get on with the other natives on the plantation well enough. Sung-in looked quite different from the locals, however, with very dark skin and a steeply raked forehead that could almost be described, albeit unkindly, as Neanderthal. He quickly learned that I liked chips so soon we were served chips with almost everything we ate. His cakes were not bad either. Margaret taught him how to make scones but forgot to tell him to divide the mixture into small pieces; Sung-in's first scones must have weighed close to a pound each.

Clothes washing was done every day by hand, using a bar of Sunlight soap; strangely, there never seemed to be any soap left over:

"Sung-in, we stap sop bilong mi?" (Sung-in, where is my soap?)

"O Missus, mi sore tru. Sop i pinis alogeta." (Oh Mrs, I am very sorry. The soap is completely gone.)

"Like hell it is," I responded when Margaret related this to me. Perhaps Sung-in was selling it on to locals; if so, we did not really mind too much as it seemed to keep him happy and was a trivial cost to us; we took to bulk ordering it from Steamships.

Each week Margaret would mail a food order to Steamships in Rabaul. It was a delight to see how quickly and competently she adopted the role of Plantation Missus and Logistics Manager. She was only 21 after all. Non-perishables arrived at Talasea a few days later by coastal steamer, while frozen food, mainly meat, was flown in on the daily DC-3 flight that landed close by at Talasea's airstrip, located on Volupai. The house was situated a few hundred metres from the western shore of the Willaumez Peninsula so we were often able to buy fresh fish from native vendors. Fruit and vegetables could be obtained locally from native garden growers. Much of what we ate came out of cans, however, and it was remarkable just how many different foodstuffs were available in canned format. Things never seen on supermarket shelves in Sydney were routinely available from Steamships. I was particularly fond of the large meat pies that came in flat, pie-shaped cans.

Food could also be purchased at Eddy's trade store in Talasea but the choice there was very limited, as he catered mainly to the native population, with canned fish and rice being the staples. What we did purchase regularly at Eddy's, however, was tobacco, which came in moist, plaited sticks about 20cm long. They had a wonderful smell, sweet, alluring, like liquorice. Sticks of tobacco were issued to our workers as a bonus and they were very well received, although once we discovered that they were using newspaper to roll their own, we were less enthusiastic. Still, they had been doing it for many years so there was no point in us trying to change their habits. We drew the line, however, at selling newspaper to them to make their cigarettes.

"Did you know that the Humphreys are importing the Saturday Sydney Morning Herald and then selling it for a penny a page to the natives?" Margaret was outraged at the exploitation.

"Sounds like a good business to me." I was quickly becoming philosophical about local mores and was anxious not to do or say anything that might prejudice our wonderful accommodation windfall.

One of the most memorable features of the house at Volupai was the row of frangipani trees that grew tall in front of the veranda. They were the largest frangipani trees we had ever seen and their intense perfume was brought into the house each evening by the gentle on-shore breeze. Remarkable as those trees were, what really charmed us most each evening as we sat out in the cool of the veranda, were the fireflies. Vast numbers of the little glowing beetles populated the frangipani trees, flashing their abdomens mostly at random but sometimes in unison. It was truly a magical site to see thousands of green flashes occurring all at once, or even, on occasion, sweeping through the trees from left to right and back again in waves of green incandescence. How these little creatures managed to communicate and coordinate their dramatic show remains a mystery to me.

There were several horses on Volupai Plantation that Richard exercised regularly when he was there. In Richard's absence, Harry suggested I use one of the horses for local transport, thereby providing it with the necessary exercise. This proposal met with a cool reception on my part. I had never really ridden a horse and the thought of controlling one of these cantankerous beasts while traversing the countryside looking at the geology was more than a little unsettling. A couple of trial rides around the plantation did nothing to assuage my concerns.

"It would be nice to ride rather than walk, but I really think I need to be closer to the ground to see the rocks more clearly." Such rationalisation seemed perfectly reasonable to me. "Besides, I am going to have climb up through the rain forest much of the time and I doubt the horse could make it up there."

"You'd be surprised where these nags can get to," Harry replied, clearly not used to having his wishes countermanded.

The clincher came when it was discovered that the horse I would be riding had developed a tropical ulcer on its leg and thus could not be ridden for some time. The horse riding idea then slipped quietly from our consciousness, face having been saved for all involved by the horse's unfortunate infection.

The house at Volupai was a superb base from which to operate but there was only so much I could get to in a day's walk from there. Later in our stay I did manage to extend my range by making a couple of overnight excursions from Volupai, walking along the coast and camping out on the beach or in the bush with my local field assistants. Margaret wisely decided not to join me on these trips and was invited instead to move across to the Humphreys' house while I was away. However, having already decided that Thelma was quite strange, she politely declined to do so. Instead, Margaret slept with the .22 rifle in the bed beside her. She would probably not have known how to use it but its presence gave her comfort and just the sight of it would have been enough to deter an intruder. Fortunately, no such event ever occurred.

As interesting as Talasea, Volupai, Voganakai and other nearby places were, I knew that my real challenge from a geological point of view was to get to the northern end of the peninsula and investigate the large caldera that dominated the geography of the region. The jungle-clad slopes of the caldera walls and the horseshoe-shaped lake inside it had been visible from the Fokker on our flight from Lae to Rabaul. The Captain had even invited me into the cockpit to get a better view and take photographs. The 1895 eruption from the new volcano that had grown inside the caldera produced lava that stood out prominently on the post-war black and white air photos I was using as a mapping base. It could be seen as a black, bare delta shape, extending from the crater at the top and spreading laterally as it flowed downhill. It was just about the only land not covered in jungle.

The Willaumez Peninsula extended for another 30km to the north from Talasea but there were no roads and walking that far was not an option. It was becoming very clear why Norm Fisher had recommended I study this long narrow peninsula: Travel by boat was really the only way to go and I began to investigate how this might be achieved.

After some discussion with Wilhelm, the Kiap, and others it emerged that a medical officer from the hospital in Talasea, Barry Hiscox, was soon to make a routine inspection visit to the villages of Bulu Muri and Bulu Dava at the northern end of the peninsula. It appeared that there would be space on the boat he was using for us to join him. Once again, the generosity and general goodwill of the expatriates in Talasea resolved a looming problem for me. It would not be the last time we were helped so materially.

Departure was set for Monday, August 7th and so most of Sunday was spent packing everything we could think of that might be needed for the two weeks we planned to be in Bulu Muri. We had precious little information about the place so it was a case of 'if in doubt, take it'. That included the 7-lb sledge hammer that I had brought along to ensure I could obtain fresh, unweathered samples from rock outcrops. Fortunately, I always had a native field assistant to carry it for me. At least we knew we would have a roof over our heads, as we were due to occupy the Bulu Muri '*haus kiap*'. Every village was required to have such a house for when the District Officer ("Kiap") came calling. It would be a good start, but having seen the *haus kiap* in some of the villages near the plantation I was nervous about what we would find in Bulu Muri.

On the Sunday night before our departure we were invited to the Humphreys' house for dinner, an invitation we gladly accepted, as we had no fresh food of our own left at that stage and just about everything else was packed for the trip. It turned out to be the most formal meal we had experienced since our wedding early in July. With lace table cloth, polished silver and fine china, the table looked a treat and the incongruity of such a formal presentation in such a remote and primitive environment did not seem to matter. In fact, making the effort to be 'civilised' was clearly part of the Humphreys' strategy for maintaining sanity in a place that severely tested the patience of the expats on a daily basis. Adding to the strangeness of that evening was the music that accompanied dinner: *Herb Alpert and the Tijauna Brass*.

"I just love the Te-jorna (*sic*) Brass, don't you?" declared Thelma.

It would have been bad form indeed to have corrected the Hostess, so I bit my tongue and swallowed bravely the green lime GI cordial she served us. They were an eccentric couple in many ways but they were well meaning and certainly very helpful to us. We had to tread carefully though. On one occasion Margaret brought the mail in from the delivery box down on the road that ran through the plantation, only to be lambasted by Thelma:

"Bringing the mail in is my job!" she exploded, as Margaret stared wide eyed in surprise at the vehemence of Thelma's reaction. Needless to say, Margaret left the mail (some of which often was for us) where it was delivered after that and directed her attempts to be helpful elsewhere. She drew the line, however, at giving Thelma a massage, as she was once requested.

In the after-dinner conversation that Sunday night I learned something of the history of Talasea and wise counsel was proffered in terms of dealing with the native population and maintaining security. Risks to personal safety were not great at that time in New Britain but we were happy to pick up a few tips from Harry. These were put to good use later when we found ourselves alone in remote native villages. It came a bit late to act on it but towards the end of the evening Harry unsettled us by saying:

"That boy Peter you've hired is bloody useless. You'll have to watch him very carefully."

"It's a pity he didn't tell us this a day or two ago," I lamented to Margaret as we walked back to our house through the night air, fragrant with aroma of flowering frangipanis.

The next day we were up before dawn, finalising our packing and preparing for the car pick up that the Kiap had arranged for us. We had decided to take Peter, who came from Volupai village, along with us to Bulu Muri as an assistant. Harry's description of Peter on Sunday night had unsettled us, but at least we knew what to expect. Bulu Muri was completely uncharted territory at that stage. Peter had been told to stay overnight Sunday night in the workers' camp on the plantation so as to be ready early on Monday. This, of course, did not happen and I was already ruing my decision to hire him without prior reference to Harry or another expat resident.

Peter still had not appeared when Barry turned up in the vehicle to transport us to the Talasea wharf. In yet another indulgence to our inexperience, he drove us around the local tracks till we finally found Peter walking in from his village. As best I could in my rudimentary Pidgin, I expressed my displeasure to Peter, who, to his credit, at least looked remorseful. But there was no time to waste so it was into

Talasea, down to the wharf and onto the motor boat that was to take us all north along the east coast of the Willaumez Peninsula to Bulu Muri. The excitement that Margaret and I felt at finally getting into the meat of my research task was tempered by a healthy dose of trepidation. The Bulus were a different tribe from the people at Talasea; how would we be received and what new challenges were about to surface?

Setting off from Volupai for a day in the field

12

BULU MURI

The Bulus, we had been told, were not from the Talasea area originally. According to legend, several generations ago they had been invited to a *Sing Sing* by the locals, arriving in canoes from the Vitu Islands to the northwest. It seems they were far from gracious guests. Clearly not satisfied with the food they were offered, they ate their hosts instead and settled down in their place, occupying what became the two northern villages of Bulu Muri and Bulu Dava. Aware of this, it was with some trepidation that we climbed down from the motor boat, anchored outside the reef, onto an outrigger canoe that had come out from the beach in front of Bulu Muri. Barry Hiscox had gone ashore ahead of us and then his wife, Olive, followed on another canoe. As we loaded ourselves and our gear onto the third outrigger, we chuckled at the outlandish sight of Olive, sitting erect on a folding aluminium chair set up on her canoe's platform, surrounded by her two children, two native women ("*meris*") and a baby, all sheltering in the shade of a large black umbrella.

The boat trip from Talasea had taken about three and a half hours, travelling through calm waters that reflected the tropical heat straight back into our faces. Never far from shore, but safely outside the fringing reef that surrounds the peninsula, we saw little sign of life on shore and carefully noted the forest-clad volcanic cones that dominated the skyline. Mapping and sampling the geology of this peninsula was clearly going to present some significant challenges.

Barry Hiscox was not pleased with the unsanitary state of the village and we watched as he called the village together and issued instructions, all in Pidgin, to rectify the situation. We had little idea of what he was actually saying, but the villagers responded promptly to his directives and there was a flurry of tidying up, sweeping the ground and shooing of pigs. The *haus kiap* that was to be our home for two weeks was built on the beach, a little separated from the village proper. It had neither shower nor useable toilet.

"*Yupela, wokim haus pek pek kwiktaim,*" ("You men, build a toilet quickly") Barry ordered.

While the Hiscoxs and Lowders watched, they with practiced frustration, we in wide-eyed wonder, the villagers hurriedly constructed a toilet shack over an existing pit near the *haus kiap*. But there was still no shower.

All too soon, Barry and Olive were ready to return to the motor boat to resume their journey around the coast to Bulu Dava, the next village, opposite Bulu Muri on the western side of the peninsula and separated from it by the caldera.

"We'll see you in Talasea in two weeks then; are you sure you'll be alright?" The tone of Olive's voice as she spoke to Margaret showed more than the obvious concern, it revealed an incredulity about what we were doing.

"Of course. We'll be fine," Margaret replied, "we're well prepared, we have plenty of food and we won't be doing anything stupid." In return, the tone in Margaret's voice betrayed a well-founded nervousness and apprehension. We really had very little idea of what to expect.

And then they were gone.

Suddenly, before lunch time, our last ties with Talasea's expat community were cut. For the next two weeks, we were on our own.

Which is not to say there were not plenty of people – everything we did was observed with keen interest by an amused gallery of villagers, who seemed to have nothing better to do than watch the young Masta and Missus set up house. But there were no other Europeans and no means of communicating with the nearest of them, 30km down the coast at Talasea. Clearly, the success of our venture was entirely in our own hands and our rudimentary Pidgin was about to get a serious workout.

First, we had to make the *haus kiap* more habitable. It was a small but typical native hut, up on stilts to stop the pigs from entering and to catch the cooling breeze that blew in off the Bismarck Sea, just metres away. Coconut palms framed the million-dollar view along the beach and out over the reef. The walls and roof of the house were made of sago palm fronds, tightly woven and draped in layers that were surprisingly waterproof. The floor consisted of bamboo slats with spaces between them, bound on to bearers with local vines. The house was divided into two parts: an open front section, a kind of veranda, that became our kitchen and dining area, and an enclosed, supposedly private back room that served as the bedroom. Privacy was relative, however, as from time to time we found small brown eyes peering up at us through the gaps between the slats. These belonged to the village children, who sat on the sand around and beneath the hut and watched everything we did; well almost everything. I did have to chase them away at certain times – we were newly married after all. Margaret was the first expat (European) woman to stay overnight in the village and as such was a great novelty for the villagers.

At that stage, we had been married just one month and if the house at Volupai was our first marital home, the *haus kiap* at Bulu Muri was the first time that Margaret was called upon to set up an empty house.

"We need a table and something to sit on," she declared with admirable authority.

"And something to sleep on," I added, speaking the obvious.

With some difficulty we managed to convey that we wanted the villagers to build a table and benches in the front part of the *haus kiap*, the open sided veranda. It is amazing how versatile bamboo and palm leaves can be in combination and the village men were really quite adept at using the natural materials.

Then it was time to make the beds.

"*Nau yu mas wokim tupela bet insait long haus.*" (Now you must make two beds inside the house.)

"*Slipim tupela diwai insait long dispela seil bilong mi.*" (Lay two poles inside this canvas of mine.)

They seemed to understand, and had probably done it before, as in no time we had two narrow beds, made by sliding poles cut from the bush through the canvas sleeves provided by Bob Heming and perching them precariously on X-shaped struts. That left only the most awkward space in the enclosed area for moving around but it would have to do.

As the day wore on, we unpacked and settled into our new abode. Cooking would take place on an open fire on the sand in front of the hut and fresh water was fetched for us from a well at the far end of the village. Primitive and not very convenient, but I heard no complaint from Margaret. Indeed, she was showing a strong pioneering spirit and a degree of unflappability that was a calming influence on me. Not many newlyweds are called upon to set up house in a hut made of palm fronds, on a tropical beach, in a remote native village, far from family and friends, with no communication to the outside world and surrounded by inquisitive but not very helpful women and children from the village, none of whom spoke English.

All but a couple of the twenty to thirty houses that constituted the village of Bulu Muri were up on a bank behind us, lined up either side of a cleared common area, parallel to the beachfront. The houses were constructed of the same local materials as the *haus kiap* but most were substantially larger than our little abode. We were glad the *haus kiap* was pretty much separate, down on the beach, as it spared us some of the unwelcome aromas that permeated the village. Pigs and chickens roamed freely throughout and it soon became clear why Barry Hiscox had reprimanded the villagers for their lack of hygiene. At least they made an effort to keep the pigs away from our place and some of the older women took to a daily sweep of the sand in front of our hut, using brooms made of stripped palm fronds. It was hard not to smile at their daily ritual. With voices raised in staccato chatter, they bent their wizened bodies forward, pendulum breasts swinging in rhythm with the vigorous sweeping.

As we were but metres from the warm waters of the Bismarck Sea we took to bathing on the beach and rinsing off under a shower bag hung from the corner of the hut. After a few days of that, the village men took pity on us and constructed a shower enclosure, comprising a circle of palm fronds stuck in the sand with a pole across the top from which to suspend our shower bag. It was sort of private. The toilet was of the long drop type, with a crude seat built over a deep pit, water somewhere at the bottom, and surrounded by a shack consisting of yet more palm fronds. Not exactly luxurious, but functional enough for a two week stay.

There was no time to waste. Much ground had to be covered over the next fourteen days and I had little idea of what to expect. Part of the time would be devoted to coastal outcrops to be sure but the main reason for the visit to Bulu Muri, and indeed, the focus for our entire sojourn at Talasea, was the large caldera at the northern end of the Willaumez Peninsula. The Dakataua caldera lurked up and over the hill behind the village: A geological Shangri-La waiting to be discovered.

Margaret and the Haus Kiap
on the beach in Bulu Muri

Baking this fish in alfoil caused great consternation in the village

The view from the Bulu Muri Haus Kiap

Olive Hiscox arriving in style at Bulu Muri

FIRE

Calderas are formed when large volcanoes collapse in on themselves. This happens when so much material (lava and ash) is ejected from the volcano that the magma chamber beneath it becomes a void. The top of the volcano then is unsupported and collapses into that void under its own weight, creating a large, crater-shaped opening at the surface. Crater Lake at the eponymous Crater Lake National Park, in Oregon, USA, is a well known example that formed 6,000 years ago. The Dakataua caldera that I studied is of similar size. At the time, I thought it was probably of comparable age because both Crater Lake and Dakataua contain similarly-sized younger volcanoes that have grown inside them as volcanic activity continued.

Since my time there, other geologists have deduced that the Dakataua caldera was formed as recently as 800 C.E. At that time, a huge eruption is thought to have occurred, ejecting around 10 cubic km of ash and lava. In size and intensity, the eruption that formed

Dakataua caldera is thought to have been comparable with the famous eruption of Krakatoa in Indonesia, which took place in 1883.

My research at Dakataua revealed that the original volcano probably rose to about 3,000 metres above sea level before it collapsed. This was inferred partly by projecting the remaining outer slopes of the volcano upwards in profile and partly by comparing the footprint of Dakataua with other PNG volcanoes that have not yet collapsed. Uluwun, also in New Britain but closer to Rabaul, is a good example. Uluwun is 2,334m high and continues to be one of the most active and dangerous volcanoes in PNG.

Dakataua volcano at sea level is roughly 18km across and the lake inside the caldera is up to 11km wide. The volcano is situated right at the end of the peninsula and remarkably the caldera wall has not been breached, which would have allowed ingress by the sea, as happened at Rabaul. The lake is as little as 450m from the sea at its closest point and is composed entirely of fresh water. Its surface is about 75m above sea level and crude depth sounding on my part, using a large lead sinker on the end of a long fishing line, indicated that the water is at least 155m deep.

13

LAKE DAKATAUA

"I speak English."

The disembodied voice drifted through the darkness, rising above the hubbub from the crowd in front of us. Then a grinning face appeared as the speaker moved in from the back of the gathering.

"Thank God," I exclaimed, extremely relieved that our impasse, it would appear, was about to be resolved.

It was the evening of our first day in Bulu Muri. We sat at the front of the *haus kiap*, where a crowd of villagers had gathered to discuss a matter of obvious importance to them but, due to our limited Pidgin, one that remained a mystery to us. Earlier that afternoon I had taken our boy Peter with me as I climbed up behind the village and over the caldera rim to take my first look at the lake that partly filled the caldera: Lake Dakataua. Together we had scrambled down the steep inner wall of the caldera to the lake shore, where we found a small outrigger canoe. For an hour or so we had paddled around the nearest part of the lake while I attempted to shoot some of the numerous ducks that populated the area, using the .22 rifle I had with me. My aim was not bad, as I managed to drop two ducks, although one disappeared without trace and the other proved to be too small to eat. It was obvious that I needed to bring the shotgun up there to make the hunting viable. After returning the canoe to where we found it, we retreated to the village luckless and duckless, as the sun sank towards the western rim of the caldera. A sense of confident anticipation flooded through me as we made our way back through the rain forest and down to the village. Studying the geology of the caldera was going to be very challenging, yes, but I felt it was all going to work – we would rise to this challenge and be successful.

Just a couple of hours later my confidence received a serious setback.

It was soon after dinner when a deputation of village men, keenly watched by women and children, came to our hut expressing concern about what I had done that afternoon (there were no secrets in this village). At that stage, the conversation was conducted entirely in Pidgin and although our language skills were improving rapidly, we were struggling to understand a key part of their concern. It was clear that they did not want me using the canoe on the lake because it belonged to a bad man, who would want a hefty payment for its use. Instead they were offering me another canoe, presently in the village, that could be transported up to the lake, where I could use it for free. That much sounded fine. The problem was that they were asking for something more:

"*Yu mas baiem bensin bilong bot.*" (You must pay for the petrol for the boat.)

"I know that '*bensin*' means petrol," I admitted to Margaret, "but what the hell has petrol got to do with a canoe on the lake?"

Again and again I asked for more explanation. Again and again they went back to asking us to pay for the petrol. Frustration and angst grew as voices were raised, heads were shaken and arguments broke out amongst the crowd. It was becoming quite heated and a just little bit scary. We had no one to call on for help, we were totally isolated and we were complete novices at dealing with the native people. I was beginning to wonder whether coming here was such a good idea.

"I'm buggered if I know what's going on here Margaret but I don't like the direction it's heading in."

"Stay calm and contain your anger," Margaret advised. "It's important to show who's in control. I'm sure we'll work it out eventually."

She was right, for it was then that the English speaker intervened. Why it took him so long to do so was never explained but perhaps he had been summoned from elsewhere in the village when it was clear that we did not fully understand their concerns.

"They don't want you to use the canoe already on the lake," he explained. "It belongs to a bad man, a murderer."

"I understand that," I interrupted, impatient but alarmed at the murderer description.

"Instead they want you to use Utu's canoe, that one over there," he pointed into the darkness, "but you must pay for the petrol that was used when that canoe was towed here from another village by the Council speedboat."

"Ah! That explains it." Relief flooded through the same veins that had carried confidence earlier in the day.

"Please tell them I will be happy to do so," I responded quickly, "just find out how much I need to pay."

I readily agreed to the small payment that was requested.

As the crowd began to disperse we quizzed out saviour about his English. He explained that he had been away to high school around the coast at Kimbe. Before he and the others all disappeared into the darkness we arranged for Utu's canoe to be taken up to the lake the next day. For 20 cents each, a group of the village men were to carry the canoe from the village beach up through the forest and down to the lake inside the caldera. Once it was there we could use it as much as we wished for transport around the lake. That night, as we prepared for bed, we were full of excitement for the day to come.

***

At 7.00 am the next morning the canoe began its journey. First the platform and outrigger had to be dismantled so that the canoe itself, a hollowed-out log about 6m long, could be man-handled through the forest. As the canoe was carried through the village to the start of the track leading up the outer slope of the caldera a throng of women, children and dogs joined the procession. It took about 10 men to lift and carry the canoe, which was simple enough on the flat land of the village. Once they began to progress through the rain forest, however, the difficulty and effort required increased commensurately with the slope of the ground. While the men called out instructions to each other, the women jabbered away in encouragement and the children ran around laughing and weaving through the procession. The village mutts also loved the action, barking furiously amidst the commotion.

Margaret and I watched on, anxious but intrigued and bemused by it all.

"What a circus!"

I could only agree, as I watched the procession with mounting trepidation.

Near the top of the slope, approaching the rim of the caldera, the ground became steeper and once or twice the canoe nearly slipped from the grasp of the straining men, causing the yelled instructions from the hangers-on to double. But if going up was difficult, the next leg of the journey looked impossible. The inside wall of the caldera was nearly vertical and the track down to the lake shore zig-zagged back and forth through scrub and forest, with barely a foothold. With bated breath Margaret and I watched as the canoe went over the rim and began its perilous journey down to the lake shore. How those men managed to control the canoe's descent without hurting either themselves or the gaggle of onlookers I will never know, but they did it. Finally, after some four hours of effort, the canoe was floating on the lake, with platform and outrigger re-attached.

***

Two men from the village had volunteered to paddle the canoe for me and act as paid guides and carriers (Peter was useless, just as Harry Humphreys had predicted). Utu, the owner of the canoe, was one of them. He was about 30 years old and unmarried. Of nuggetty build, as strong as a horse and very agile, he became a faithful and hard-working assistant for the remainder of that visit to Bulu Muri and again on subsequent visits. It was he that willingly carried my 7-lb sledge hammer everywhere we went and he showed a lot interest in what I was doing. His footwear caused us much amusement though – two thongs, one yellow and one green, both designed for the left foot.

The other man we initially called Lari, although we discovered much later that his name was actually Gari, similar to my own. He was older, around 45, taller and leaner than Utu but equally strong and hard-working. A red cap was always on his head, a beacon in the dim light of the rain forest. The father of four children (*"fuwapela pikinini"*), he was quite a humorist, forever laughing and joking, in contrast to the more taciturn Utu. Gari was quick to offer suggestions and proved most resourceful as time went on.

These two men quickly became more than paid labourers, as they settled into their roles and became my companions and friends. My Pidgin improved rapidly as we conversed and I asked them about their families and the history of the area. Both were anxious for our welfare and Gari in particular adopted an almost fatherly attitude towards us, which, as naïve 21 and 23 year-olds, we really appreciated.

"I think the natives are much easier to get on with than the expats," I commented to Margaret after about a week in Bulu Muri.

"Well they're certainly not troppo," she agreed, "and they seem much better adjusted and at ease with their world."

"Well is it any wonder? Look at where they live."

As I spoke, a group of naked children frolicked on the beach, laughing and calling out in glee as they jumped into the warm waters of the Bismarck Sea.

WATER

"*Pul long dispela hap*," (Paddle in this direction) I requested, as we set off in Utu's canoe to begin my exploration of the caldera from the inside. Utu sat in the bow ("*long hap i kam*") paddling hard, while Gari was in the stern ("*long hap i go*") tirelessly providing both propulsion and steering. The latter was no mean feat, given the incessant drag of the outrigger. First up was an excursion along the eastern shore of the lake. To be honest, I felt rather grand sitting on the canoe's platform, my feet resting inside the dugout while Utu paddled strongly in front of me. Margaret had remained in the village for this first day.

Our route took us south, along the eastern shore of Lake Dakataua. My eyes were turned towards the shore, examining the rock outcrops we were passing. For much of the way the inside wall of the caldera was close to vertical and climbing up or down would have been dangerous if not impossible. Even so, stunted trees dripping with vines and creepers imparted a green hue to the slopes and masked the higher outcrops. Clearly, plant life here was tenacious and wherever a modicum of soil could accumulate something grew.

"*Waitim mi hia pastaim*," (Wait for me here for a while) came my frequent instruction, as I stepped ashore when local topography permitted. Energetically, I wielded the sledge hammer to take samples, recording details in my field note book.

After a few hours of this we rounded a prominent bend in the caldera wall and noticed a narrow cleft reaching from the lake surface high up into the wall in front of us. Several large bats were emerging from the cave, flying over our heads then retreating out of sight back into the cave.

The little boy in me took control. Just as he might have trampled on an ants' nest to disturb them and bring them out to attack, before stepping safely backwards, he felt driven to provoke the bats to see what would happen.

Under his spell, I picked up the shotgun that rested on the platform behind me and fired a round into the cave.

Suddenly, the sky was filled with bats, screeching and flapping over our heads. The little boy hastily withdrew.

"Bloody hell, there must be thousands of them in there!" I exclaimed aloud, the hair on the back of my neck standing on end.

The canoe rocked with my excitement but my companions remained calm, as Gari made a polite but firm request.

"*Masta, yu mas siutim sumpela liklik blakboks. Em i gutpela kaikai.*" (Mister, you must shoot some bats. They are good eating.)

I settled down quickly, relishing the chance to show my expertise with the shotgun now that so many bats were airborne. Reloading, I aimed into the billowing cloud of furry creatures and pulled the trigger. Three or four dropped into the water. As we paddled over to collect them the remaining bats went berserk.

"Hey, the bastards are shitting on me!"

What did I expect?

"Let's get out of here! *Pul kwiktaim*."

Tonight's *kaikai* safely on board (for Gari and Utu; I politely declined to share the spoils), we rapidly turned the canoe around and discreetly departed from the chaos.

***

With time still to spare before dark, I requested Gari and Utu to take us back past our starting point to look at the outcrops to the north. It was there that we had an encounter with another of the forest's wild inhabitants. As we had done already many times that day, I called for a stop to sample some interesting rocks I could see in outcrop. Gently the canoe nudged into the shore and Gari, who on this occasion was paddling up front, stepped ashore to secure the canoe. Within seconds he was back on the canoe, climbing past me on the platform and nestling up against Utu at the rear. As he did so he muttered guttural sounds and wrapped his arms around himself, obviously frightened. Anxious to determine the cause of this display I stepped ashore myself. Then I saw it. A giant python was writhing over the rock seeking cover from us intruders. I never did see the head but I estimated that the body I saw was at least 5m long and as thick as my leg. I returned to the canoe as we hastily withdrew.

Our next encounter with a reptile was rather different, as I was treated to the spectacle of a lizard capture. I had already acquired, at the cost of a can of condensed milk, a *'kundu'* – a hollow wooden drum – but it lacked a skin for the top. So, when Utu saw a suitable lizard climb a tree overhanging the edge of the lake he set out to capture it so he could use its skin to cover my drum. Stepping ashore, he quickly cut a long pole from a sapling in the bush, together with a length of vine. Deftly, he tied the vine into a loop at the end of the pole and trailed the rest of the vine down to his free hand. As the lizard sat unmoving on a branch Utu stood in the canoe and cautiously raised the pole. Very slowly he slid the loop over the neck of the lizard and with a sharp tug the loop closed and the lizard was caught by the vine and pulled down. Without hesitation Utu grabbed the half metre long lizard by the neck and proceeded to break each of its legs before throwing it into the bottom of the canoe. I felt bad at being the cause of such dispassionate cruelty, but I knew that this was ordinary business to these people of the forest. To this day, I have that *kundu*, still with its skin attached, although the sap beads ("*susu*" – nipples) that Utu stuck onto the skin have long since fallen off.

***

Two days later our exploration took us across two kilometres of open water to the shore of the new volcano in the centre of the lake. While the growth here was just as verdant, the shoreline was much gentler, with low rocky bars and sandy beaches, many of them overhung with pandanus trees and other greenery. At one point, as we paddled slowly along the shore, a large splashing sound came from beneath the overhanging shrubbery.

"*Pukpuk!*" exclaimed Utu.

My tendons tightened as I scanned the water beside us. Despite their freshness, the waters of Lake Dakataua host numerous saltwater crocodiles. On this occasion, the beast slinked away from us, seen only by Gari as a murky shape beneath the surface. How these animals got into the lake in the first place remains a mystery.

"How did they know there was a lake up here?" I pondered as we kept a keen eye open for more reptiles.

Then it was time to go ashore and make our trek to the top of Mt Makalia, the little volcano (350m above sea level) that rose above the waters in the centre of the lake. It was this volcano that erupted in 1895, spilling a black lava flow down its eastern flank. To the south, overlapping ash cones and other volcanic deposits formed a small peninsula that joined Makalia to the southern rim of the caldera. The ascent took an hour or so, with Gari in the lead and Utu carrying my gear, including the 7-lb sledge hammer. There was no track but Gari seemed to know where to go as he slashed at the undergrowth with his machete, cutting a relatively clear path for me as I followed behind.

"*Lukaut Masta. Mong mong i stap*," came Gari's frequent warning, as we passed a wide variety of stinging plants. Some of these looked threatening, their broad leaves full of holes, with nasty hairs visible on the underside. Others appeared quite benign and I was glad to have Gari's local knowledge to stop me being stung unnecessarily.

As we made our way up the slope through the forest I was startled to hear a roaring sound, like some angry beast. The adult me knew there were no dragons in the bush but the little boy surfaced again, nervous about what it was that made so much noise.

"*Wonem meknais olosem?*" (What's making such a noise?) I asked.

"*Bikpela rein i kam*," (Heavy rain is coming) replied Gari, as he cut a metre-wide leaf from the undergrowth and offered it to me as shelter. Minutes later the light dimmed dramatically as the downpour burst upon us. It was the roar of Niagara Falls right over my head, drowning all other sensory experience. Cowed by the intensity of the torrent as it struck the canopy above us, I crouched beneath my leafy umbrella and waited for the rain to abate.

By the time we reached the summit and broke out into the clear, where the 1895 lava flow remained largely free of vegetation, the rain had stopped and the sun was again shining. The lava had issued from a crater at the top of Makalia and flowed out through a breach on the eastern side. The surface of the lava was a jumble of black, brittle rocks with crumbly surfaces that made walking over the flow hazardous and slow, especially for Utu, in his two left thongs.

"*Waitim mi hia pastaim*," (Wait for me here for a while,) I told Utu, as we stood on stable ground at the edge of the flow. Equipped with sturdy field boots, I picked up the sledge hammer and walked hesitantly out onto the lava, the blocks crunching and grinding together under my weight. Gari, wearing old gym boots, followed cautiously. Swinging the sledge with gusto, I started collecting samples and soon had several specimens of fresh young lava – a mere 72 years old. Utu, meanwhile, was examining a spot inside the crater where steam and volcanic gases had reacted with the lava and converted it to vermillion coloured clay, some of which he collected for later ceremonial use. The colour, I reflected afterwards, was probably due to traces of the mineral cinnabar – mercury sulphide – which is a common product of volcanic fumaroles. Wisps of steam continued to drift up from gaps between the rocks inside the crater and the by-now-familiar aroma of hydrogen sulphide gave the scene a surreal atmosphere.

"*Bihain, mi laik bringim Missus bilong mi long dispela hap*," (Later, I would like to bring my wife to this place) I informed my guides, as they put my samples into the backpack and prepared for the journey home. Then it was back down to the canoe, across the lake, up the steep inner wall of the caldera and down the relatively gentle outer slope to the village on the sea shore.

Margaret was sitting on the steps of the *haus kiap* chatting to a group of women as I approached.

"Guess what!" she called, "we've got lobster for dinner!"
"Beats bats!" I called back.

Lake Dakatuna, 11km across, inside the caldera at the northern end of the Willaumez Peninsula

The ancestral mountain that once stood here was about 3,000m high and Mt Makalia, in the middle of the lake, represents resurgent volcanism

14

MAKALIA CONQUERED

It was not just for dinner that we ate lobster. Once our taste for the local crayfish became known the villagers went hunting for "*kindam*" out on the reef and brought us several more, although due to the market demand we had created, the price of each one went from 10c to 20c overnight. Still, 20c for a lobster breakfast was hard to beat. Fish and other fresh foods were also offered to us for purchase. The vegetables were generally fine but we drew the line at eggs, as most were fertilised and many were well on their way to becoming mini fumaroles – small packages of rotten egg gas (hydrogen sulphide).

Our cooking was carried out over an open fire in front of the *haus kiap*. Margaret quickly became adept at producing tasty meals from a mixture of cans, packets and local fresh foods. She even managed to make bread and pancakes. One memorable meal took place after I had caught a large fish, of unknown identity but something like a trevally, while canoeing over the reef. Up till then we had simply fried fillets of the fish either caught or bought. This time, given its bulk, we decided to bake the catch in aluminium foil, marinated in coconut milk. Several village women looked on with scepticism, tut tutting as I placed the package amongst the coals of the fire.

"*Em i longlong,*" (He's crazy) they muttered, as I attempted to cook the fish in, as they thought, paper; aluminium foil was a totally new experience for them.

"*Nogat. Weitim pastaim na lukim yu.*" (No. Wait a bit and you'll see.)

They were stunned and very excited when eventually I took the foil package off the fire and opened it up: A beautifully baked whole fish looked at them through its glazed eye as the aroma of hot coconut milk wafted upwards. Other women were called to come and view the magic of our shiny cooking "paper".

It was not easy to roast food on the fire though, with or without foil, as the wood we burned was all quite soft and generated few coals over which to bake anything. Each meal consumed a substantial amount of firewood but fortunately the village men kept us well supplied. Another item supplied regularly was what the locals called a *kulau*, which was a green coconut fetched from a tree and slashed open at the top with a machete. The sweet, slightly effervescent liquid inside was always refreshing and a welcome drink in our unrefrigerated tropical state. At the end of each day it became my clarion call:

"*Kisim kulau i kam.*" (Fetch a green coconut.)

It was always entertaining to see the skill with which the young men scaled a tall coconut palm by wrapping a piece of vine in a loop around the tree and their feet and then using that as a slipping anchor to propel themselves up the trunk, machete tied to their waists or even sometimes between their teeth. At the top, they then deftly slashed off a green nut, which fell to the ground with a resounding thud. Coconuts fall from the trees by themselves when ripe and represent a significant hazard in tropical regions; we noticed that few people sat directly under trees with yellowing (ripening) coconuts.

With no refrigeration, keeping food was a problem. Generally, fresh foods were either consumed the same day or discarded, although fruit usually lasted longer. One morning though, as I was getting dressed, Margaret called from the veranda:

"Garreeeee! Something's eaten some of our bananas."

"Rats!"

This was not an expostulation but a statement of fact – we had clearly suffered a vermin attack overnight.

"Ugh!" Margaret had been very resilient in the face of all the challenges our stay in Bulu Muri had thrown at her but this was the last straw. I was equally disgusted.

Our distress was all the greater because we had spent an uncomfortable night after fleas invaded our beds and kept us awake. We found later that washing the sheets daily mitigated the flea problem, but it was risky, as it rained every day and getting our only sheets dry before or after the rain became another challenge for Margaret. At least we were spared the irritation of the sand flies of Talasea and there were very few mosquitoes.

Malaria was widespread in PNG at the time, although it was not a major issue in the Talasea area. Nevertheless, we took our antimalarial drugs regularly and neither of us had any problem with that disease. We did suffer some other illness though, principally diarrhoea – with no refrigeration, good food hygiene was difficult to maintain in the primitive setting of a tropical *haus kiap*. My main medical issue though was stinging plants. While Utu and Gari did their best to point out the *"mong mong"* plants to me, inevitably, I brushed against some of them and suffered the consequences. On one occasion, later in our stay at Talasea, I came home to Volupai with my right thigh covered in a red rash it was impossible not to scratch. My leg swelled up and the mottled red skin became as tough as leather for a few days then slowly recovered; I never did find out exactly what had caused this reaction.

It became our pattern while in Bulu Muri that Margaret accompanied me into the bush about every second day and remained in the village on the intervening days. Her days at home were spent reading and chatting to the women of the village. As her Pidgin improved the conversations grew longer and more complex. One of the first things the women wanted to know was:

"*Yu gat pikinini Missus?*" (Do you have children Mrs?)

"*Nogat.*" (No.)

"*Haumas taim yu bin Missus?*" (How long have you been a Mrs?)

"*Wanpela mun.*" (One month.)

By Margaret's account the uproarious laughter that followed this answer ran in audible relay from the women on the beach up to those in the village nearby and along to the farthest house. Just why the revelation of our newlywed status caused such amusement to the village women was unclear, but it was not such a bad thing. The village people soon became even more friendly and solicitous of our welfare.

My study of the geology of this area required me to examine and sample rock outcrops wherever I could. In the rain forest this was commonly difficult, as such rocks as did crop out were generally hidden by undergrowth or covered in moss and the volcanic features were obscured. Other than inside the caldera, the best outcrops were along the coast, where lava flows and ash deposits emerged from under the trees and formed little cliffs and buttresses at the shore line. With Utu's canoe up on the lake, Gari offered us the use of his canoe, which was similar to Utu's but newer and in much better condition – it did not leak for a start.

One memorable day we set off in Gari's canoe to travel around the coast to the northwest. After paddling a short distance, Gari put up a crude sail. Quickly the canoe began powering over the reef before a stiff breeze, heavy with tropical salt. It was pure exhilaration. Ahead of us the water was brilliantly clear, allowing the sunlight to penetrate deeply, where it seemed to be broken into its components by the reef and reflect back into our faces with a multi-coloured, animated sparkle. The gentle swish of the sea water against the canoe as we sailed along was a warm, soothing sound, making us feel as much at ease in this place as Gari and Utu obviously were. Brightly coloured fish could be seen darting amongst the coral as we peered into the water around us. It was pure bliss. Relaxed as we were, it came as a shock when Utu suddenly jumped off the canoe and disappeared under water. Seconds later he emerged holding a turtle about half a metre across and tossed it into the canoe.

"*Tarasel gutpela kaikai*," (Turtle is good eating) announced Gari as Utu, looking very pleased with himself, clambered back into the canoe. He and Gari chatted away for several minutes, presumably working out how they would share the chelonian spoils.

At regular intervals we pulled into the shore so I could examine the rocks and take some samples. While I did that, Margaret sat on the coral sand under overhanging trees, seeking relief from the intensity of the tropical sun.

"I bet there'd be plenty of people who would pay a lot of money for this experience," she observed, as handfuls of white coral sand trickled through her fingers.

"Yeah, I guess you're right. Why don't you stay here for a bit while I go on further with Gari and Utu?" I suggested.

"This beach is the nicest one we've seen and you should be fine here in the shade. That OK with you?"

"Of course."

Leaving Margaret with some water, a book and my .22 rifle, we three men continued our journey around the coast. It was a good thing that she had opted to wait there, as before long the reef became impassable by canoe and we had to beach it and walk along the coast, climbing over sharp black outcrops of lava that had flowed to the sea. Some three hours later we returned, tired and sweaty to Margaret's beach. Utu asked to borrow the rifle to go hunting and I agreed as I had come to trust this nuggety bundle of energy. Utu said he would walk back to the village with his catch. Gari, meanwhile, was happy to sit on his canoe and have a smoke.

"I've never seen colours quite like these," Margaret declared, looking out over the crenulated, sparkling water, "it all looks so inviting."

"Well then let's go for a swim," I suggested.

The lure of the pastel blue rippling water, which contrasted with the mottled browns and dappled yellows of the submerged reef, could no longer be resisted, so Margaret and I moved along the beach a discreet distance from Gari. There we had a swim that I found gloriously refreshing after so much exertion.

As we part sailed and part paddled back to Bulu Muri, with me up front because Utu had gone hunting, I trailed a fishing line out the back of the canoe and was delighted to catch our dinner.

"It looks a bit like a rock cod," I suggested, as I threw the flapping creature into the canoe and asked Gari if it was good to eat.

"*Yes Masta, em gutpela kaikai tru.*"

Later, a fully satisfied Masta announced to the Missus cook:

"Gari was right. That fish really was delicious."

Well after dark, Utu returned to the village, proudly displaying a pig and a big lizard he had shot with my gun.

***

After a little more than a week in Bulu Muri, Margaret decided that it was time for her to make the trek to the top of Mt Makalia.

"I agree. I want you to see it but you'll find it quite challenging."

"I'm getting a bit used to challenges."

The trek would involve a hike up the outer slope of the caldera and down the steep inner wall to the lake shore, followed by a two kilometre crossing of the lake in Utu's canoe. Then would follow a climb through the rain forest up the 300m or so to the craters at the top of the volcano, source of the 'black rock' 1895 lava flow.

At 8.00 am Gari and Utu came to collect us and our field gear, ready for the assault. The first part was relatively easy, as there was a well made track up to the rim of the caldera and, walking steadily, we made it in good time. The scramble down to the lake shore was more difficult but fortunately brief and before long we were loading ourselves, my sledge hammer and backpack and the trusty shotgun on to Utu's canoe.

"The paddle across to the shore in the middle should take an hour and a bit," I informed Margaret, who was showing no signs of nervousness, even though I had told her about the crocodiles in the lake.

I realise now that Utu's flimsy canoe would have been no match for a hungry 5m saltie, but back then we were still indestructible so we did not think too much about the danger. There was little wind so the trip across the lake would be smooth and our guides should be able to paddle directly to the little beach that would be our launching pad for the climb to the top of Makalia. The sky was cloudy but there was no rain about, which added to our confidence as we set off. About half way across I called a halt while I dropped my primitive sounding line to measure the depth of the lake. It sank and sank and sank, finally stopping at 155m. That day Margaret was my scribe:

"Please make a note of that depth in my field book," I asked.

"No problem," came her ready reply, but then,

"Oops!"

"What's up?"

"I dropped your pen into the water."

"Oh no, damn it!"

Fortunately, I had another with me but the lost one was my favourite, a gold Parker she had given me for my 21st birthday.

"Ah well, at least we know that it is going to rest under 155m of water until one day a lava flow or ash fall will cover it. It might even reappear as a fossil in a few million years. That will make a nice puzzle for the geologists of the day."

The paddling resumed and we soon landed on the shore of Mt Makalia.

There was no clear track to the top of the mountain but having been up there just days earlier Gari's slashed undergrowth from that excursion was still distinct enough. Around the middle of the day we set off, just as the clouds began to disperse, allowing the intense tropical sun to beat down on the forest. The forest responded to the full sun by increasing its transpiration, raising the humidity considerably. After just a few hundred metres of walking Margaret and I were sodden as sweat trickled through our eyebrows and ran in rivulets from our noses. I was more or less used to it but Margaret found the experience quite enervating. Gari and Utu seemed as immune to the humidity as the cicadas that sang around us. Then the going got steeper and the ground more uneven, as we clambered over lava flows that were geologically young but had been around long enough for the forest to reclaim the land. Margaret began to struggle with the conditions.

"How far have we got to go?" she asked, pleadingly.

I enquired of Gari.

"*Em i longwei liklik Missus,*" was his reply.

We both knew that '*liklik*' meant 'little' so Margaret was consoled to think there was not far to go. In fact, though, I knew that Gari's answer really meant: "It's quite a little way yet Mrs." I decided it was best for me to keep my mouth shut as we trudged on. At least it was not raining like it had been on my first trip to Makalia.

"*Vulkan i klostu Missus,*" (The volcano is nearby) Gari finally announced, as the forest thinned and we broke out into bright sunshine. We had made it and more to the point, Margaret had made it to the focal point of my whole field programme at Talasea. My study included most of the Willaumez Peninsula but the caldera around us was the core element of my field research. Standing there, atop this young volcano, gave both of us a sense of achievement, an objective attained, a journey ended.

"It is really quite beautiful up here," she observed, breathing heavily, as we stood on the rim of the largest Makalia crater and gazed at the lake below us, framed by the caldera wall.

"Except for the smell that is," she added, as wisps of rotten egg gas wafted past.

In every direction except immediately south of us the lake waters abutted against a near vertical wall of green-draped rock – the fault surface along which the original volcano had collapsed to create the caldera. To our north lay Benda, a block of the primary volcano that had not fully collapsed and now stood proud as an unscaleable table-topped hill. Across to our west was Doko, another residual block of the original mountain, with rocky sides that dropped sheer into the lake water. The awesome power of nature that the caldera represented was difficult to comprehend and my quiet observation seemed puny:

"It must have been one hell of a dramatic and fiery show when all this happened."

***

I tried to imagine the scene right there where we stood, just yesterday in geological terms[20]:

- An ash column rising 15km into the atmosphere;
- Lava pouring down the outside of the mountain to be quenched in the boiling sea;
- Phreatic (steam) explosions scattering the lava fragments back onto the land;
- Earthquakes rocking the ground every few minutes;
- A mighty blast as the magma chamber finally emptied;
- A roar that could be heard in Singapore (had it existed);
- The mountain collapsing in on itself to create the caldera;
- Tsunamis spreading out across the Bismark Sea to inundate the nearby islands.

"I wonder if there was anyone here to witness such a dramatic event," I pondered, "and to fall victim to the catastrophe."

After an hour or so wandering around the crater rim, crunching over brittle lava and gravelly ash, it was time to start the return journey. Carefully avoiding mong mongs, we traipsed back down through the forest, accompanied again by a chorus of cicadas. A couple of times some pigs snuffled in the bush near us but remained out of sight. As we boarded the canoe to cross back over the lake I was a little disconcerted to find that the breeze had strengthened and the water

---

[20] More recent research has dated the last big eruption at Dakataua as occurring around 800 C.E., when as much as 10 cubic km of material was ejected as ash and lava.

was getting quite choppy. Still, there was no choice in the matter so off we went, relying again on the strength and resilience of our guides. My uneasiness grew as we made our way across, not helped by the fact that the canoe had a leak where the skin of the dugout had cracked. I was kept busy bailing as Gari and Utu paddled.

Already tired, Margaret found the climb up the inner wall of the caldera quite hard but she persevered and we finally made it back to our little *haus kiap* on the beach. At my insistence, Margaret rested the next day in Bulu Muri while I returned to the lake to visit and sample the bottom end of the 'black rock' lava flow.

As the end of our two week stay in the village drew near I made several more excursions along the coast, up into the rain forest and along the lake shore to make certain I had a representative collection of all the rock types I had seen. Everywhere I went I was confronted by the raw force of nature, whether it be the volcano and its products or the forest and its irrepressible green vitality. At one rocky outcrop Gari stopped and reached into a cleft, pulling out a human skull.

"*Papa bilong mi*," he proudly announced, with a little giggle.

As we walked on I quizzed Gari about what it was like during the Japanese occupation, just over 20 years beforehand. He became uncharacteristically quiet as he explained how bad things had been, especially towards the end of the war, when the Japanese troops had run out of food.

"*Siapan solja katim susu bilong ol meris, tasol i laikim kaikai olsem.*" The Japanese had, he claimed, resorted to cutting off the breasts of women in order to eat them.

Appalled and incredulous I looked into Gari's eyes, where I saw no hint of deceit. He had no reason to make this up and was old enough to have seen it himself, although I suppose he may have been repeating in good faith a local legend. Whether true or not, I could see that he had a very dim view of the Japanese.

***

On Monday, 21st August, our two week visit to Bulu Muri came to an end. Right on schedule, the Local Government Council speedboat – in fact a 5m aluminium dinghy with a 30hp outboard motor – arrived at the beach adjacent to our house. As we gathered our gear together, including my precious rock samples, a crowd of 40 or so gathered to watch. In steady rain, we packed our goods into the boat, assisted by the villagers, and were then given a rousing send-off by the assembly. We had clearly made quite an impression on the Bulus, and they in turn on us. We carried with us money from some of them with which to buy rice to bring back to the village on our next visit. The rain cleared as we motored southwards and the sea was calm so I trailed my 250lb breaking strain fishing line behind the boat.

To my great surprise, I hauled in a Spanish mackerel about a metre long and weighing probably 15kg. Three hours after leaving Bulu Muri we arrived in Talasea and I paid the boat driver for its use – $6 for the round trip – and gave him a liberal portion of the mackerel.

As we were transported back to Volupai in the trailer pulled by the Council tractor I reflected on what lay ahead.

"The next time we go up there we will have Ian Carmichael with us," I said to Margaret. "I wonder what he'll make of it all."

I was looking forward to the visit by my supervising professor and knew what to expect, having spent the previous ten months with him in Berkeley. Margaret, not having been to Berkeley as yet, could only imagine what this enigmatic character I had talked endlessly about would be like.

He was due in two weeks.

Margaret sets out across Lake Dakataua with Gari and Utu, our precious "kago" and just a few inches of freeboard

Utu, Peter and Gari, with Utu's canoe on the lake

15

THE PROF

Out on the grassy landing strip the TAA DC-3 from Lae taxied in, its nose high in the air and its tail wagging as the pilot veered left and right so he could see where he was going. Soon it was parked alongside the shed that served as Talsaea's terminal. Margaret and I stood in the sun, fidgeting in nervous anticipation. Ian Carmichael, my PhD supervisor from Berkeley was on board, making his first visit to Papua New Guinea. Bad weather had prevented the aircraft's departure from Lae the previous day so Ian had spent an unscheduled and probably uncomfortable night in what was then very much a frontier town. I knew he would have had a long and circuitous trip just getting to Talsaea from his previous stop in East Africa, so I was expecting the professorial patience to have worn a bit thin by then.

Margaret, however, was excited and looking forward to hosting the visitor, the first of our married life. I was keen to show him the geological treasures I had thus far unearthed at Talasea. Up to that point her knowledge of the man was limited to my descriptions drawn from exposure to him over the ten months I had already been in Berkeley (when Margaret was not with me). Carmichael was a difficult man to describe to those that had not met him, but "larger-than-life" covered it fairly well.

"He's very direct and says exactly what he thinks," I had said to Margaret earlier, "and he does not suffer fools gladly. You'll have to get used to his, ahhh … colourful language."

"But if he's critical it will be to your face, never behind your back. And he can be surprisingly engaging and thoughtful when he wants to be. In fact, even when he's sounding off at you there's a hint of humour in his voice and a bit of a glint in his eye. I think much of the bluster is just part of the image he likes to present."

Margaret was intrigued but did not find my comments all that reassuring.

"I'm not sure if I'm going to like him or not. He sounds a bit bombastic and intimidating," she had concluded.

"He is," I had replied, "but he's also the most stimulating and inspiring person I have ever met."

"Hmmm. We'll see."

We were both wondering just what the next three weeks would bring and, in my case, the sweat beads forming on my brow as I waited may have been as much from anxiety as from the mid-morning tropical sun. My Prof was not noted for his patience and I feared the consequences of the unexpected stopover in Lae.

The aircraft door opened and passengers began to emerge, among them a man impressive in both stature and bearing.

"That's him!"

As we approached Ian a broad grin broke out on his face and he held out his hand to Margaret.

"Hello lass. Delighted to meet you. What a beautiful place you have here!"

No sign of a foul mood and Margaret was instantly charmed.

He then turned to me as I greeted him with pleasure and relief:

"Welcome to Talasea."

"It's taken me nine fucking days to get here from Kenya," came his quick and rather less charming reply.

"It's a bloody long way so I hope you've found something interesting."

"I don't think you'll be disappointed."

We watched as he swivelled on the spot, taking in his new surroundings.

"Jesus, what a place!"

I felt then that the visit would be a success. He was clearly as captivated by the tropical verdure as we were and he could easily see, as I did on my first day there, signs of volcanic activity all around us. Larger-than-life Ian was not slow to express his admiration for the scene and we both soon found his enthusiasm infectious. I need not have worried, because, as I would discover again and again over the next three years, whatever else Ian Carmichael might be, he was intensely loyal to his students.

"Come and have some lunch while I tell you about what I've done so far," I suggested, "and we can make plans for the rest of your visit."

"Show me which is your baggage and I'll arrange for it to be collected and brought to the house."

Ian indicated which suitcase was his and I issued instructions to Sung-in:

"*Karim kago bilong Masta Ian long haus bilong mipela.*" (Take Mr Ian's baggage to our house.)

"Bloody hell! What did you just say?" Ian looked impressed as I showed off my expertise in Pidgin.

As we were about to leave he took delight in photographing some native children and their mothers, intrigued by the fair hair they sported.

"It comes out of a bottle," I explained. "Bleaching the hair is very popular with the women and kids here."

***

Over the next week, I began to show Ian something of the local geology, including outcrops along the coast near Voganakai and Volupai villages and along the road to Talasea from Volupai Plantation.

"I see you've got some rhyolites then," he observed, as we walked back along the road from Talasea and crossed a black, glassy obsidian outcrop.

It had not really registered with me that the obsidian I had seen previously was in fact a form of rhyolite lava, so, rather than show my ignorance, I just casually agreed:

"Yeah, plenty of that."

His previous stop had been with another of his graduate students, Frank Brown, who was working with the Louis Leakey team investigating human origins at Olduvai Gorge in Tanzania. Frank was dating the fossiliferous horizons using isotopic geochronology on the intercalated volcanic ash units.

"How's Frank going?" I enquired politely.

"Very well. Fits right in with the Leakey team. But you know Frank, he doesn't let you into his thinking very much. Still, he's in his element there. Practically gone native."

All of which made me worry about what Ian would say about me after this visit.

"How does the Talasea climate compare with that in East Africa?"

"It's cooler here than where Frank is but it's a hell of a lot more humid."

"At Olduvai, everyone drives round in Landrovers but you, you bastard, are making me walk everywhere."

"Never mind Ian," I was quick to reply, "next week I'll be taking you on a delightful cruise up the tropical coast to Bulu Muri. You'll love it."

"I'd better."

***

When we took Ian to the Talasea Club, with its expansive view across the harbour, he was lavish in his praise of the natural beauty and could not understand how Talasea had not already become a tourist magnet. After I reminded him of the difficulty he had encountered just getting to Talasea he acknowledged that perhaps the Territory did need a bit more infrastructure development before the tourists could come. To the other expats at the Club he was a true celebrity – never before had a professor from America been to Talasea and they flocked to him like the moths that crawled over the lights on the veranda. Responding to their adulation, Ian turned on the charm and I marvelled at the man's capacity to flatter.

Our investigation of the district's geology continued. Next up was a two-day trek that began with canoe transport to Voganakai, north along the west coast, and included a traverse across the peninsula to the large geothermal field near the village of Pangalu on the east coast, where we camped in the *haus kiap* overnight. Pangalu lies on the north side of Garua Harbour, opposite Talasea, from where columns of condensed steam can be seen rising above the many boiling pools and mud pots behind the village. The largest pool was at least a hectare in area and was surrounded by hot mud and fuming holes in the ground, making walking near it quite treacherous. Pangalu thermal field includes several geysers, one of which in particular captured Ian's attention.

"That geyser is more impressive than the original one at Geysir in Iceland," he told me.

Ian knew Iceland well, having studied the Thingmuli volcano there for his own PhD, undertaken at Imperial College in London. I had arranged for the council boat to come over and pick us up and while we waited he regaled me with stories from his time in Iceland. The conversation continued but became increasingly strained as we waited, and waited, and waited for the council boat. Finally, in complete frustration, we persuaded a couple of the village men to paddle us across to Talasea in their canoe. Ian was not pleased, as I tried to explain the failure of my best-laid plans by reference to the "New Guinea Factor". I was sincerely hoping that the preparations I had made for our forthcoming return to Bulu Muri with Ian would be more successful.

But of course, it was not to be. On Monday, 11 September we were all set to make a week-long trip to the village, starting with a pick-up by the car from Kilu village that was due at Volupai at 7.30am. At 9.30, with no sign of the Kilu car, we prevailed upon the Humphreys in the main house to ring Eddy (the Chinese trade store owner) in Talasea. He then came out to collect us and our gear. Fortunately, the council boat was still waiting for us and we finally got away in the heavily laden boat at noon, arriving in Bulu Muri at 4.00pm. Once there, the frustrations of the day were quickly subsumed by the warmth and enthusiasm of the welcome we received as we came ashore next to the *haus kiap*. Gari and Utu stood proudly in front of the crowd and were all smiles as we landed.

"*Apinum Masta; apinum Missus,*" (Good afternoon Mr; good afternoon Mrs) came their greeting, as I introduced Ian to them.

***

Over the next week, I took the Prof on a tour of the highlights of the Dakataua Caldera and its surrounds, including a couple of nights camped on the shores of Mt Makalia, inside the caldera. On the first day in that camp, while Gari and Utu returned to the other shore to collect the balance of our gear, Ian and I set off to inspect the bottom end of the 1895 lava flow that my air photos showed to be only a kilometre or so distant through the forest. About two hours later, two hot, sweaty and grumpy men staggered back into the camp, almost by accident. We had become completely disoriented in the thick bush and failed to find any trace of the young lava flow.

"Well, well; I thought geologists could find their way around."

The geologists growled at Margaret's sarcasm.

"Perhaps you'd better wait for Gari and Utu to return and let them show you the way."

Which they did, shortly afterwards; under their guidance we reached the lava flow in fifteen minutes, much to my embarrassment. I suffered Ian's ribbing in silence. For the most part, though, Ian was generous in his praise for what I had already achieved and offered much good advice about what I should concentrate on before leaving the area.

"I can't get over how beautiful the whole place is," he kept repeating, "and I'm staggered at the density of the vegetation. How the hell do things grow on a slope like that?" He pointed to the cliff face of Doko, which was draped in solid green despite its near vertical aspect.

"I'm so pleased the geology has turned out to be as interesting as it is. Norm Fisher certainly knew what he was doing when he recommended this place to you."

Over the next two days Ian and I made excursions in Utu's canoe (with Utu and Gari paddling) around both the east and west coasts of the Makalia shoreline and climbed to the top of Mt Makalia itself. Margaret stayed in camp and prepared our meals, using only the most basic of facilities.

After two days it was time to move on. As I supervised the pack up for the return to Bulu Muri, I was startled by a verbal explosion nearby.

"Jesus Christ! Will you look at this?"

Ian stood by the water's edge, staring in disbelief at a rope, with a large hook attached, around which the intestines of a pig were wound. Utu was looking at him in bewilderment, unable to see the problem.

It turned out that the whole time we had been camped on the lake shore Utu had set out this bait line, hoping to catch a crocodile. The porcine bait had not been taken but then, fortunately, neither had the human bait that had been sleeping peacefully under a tent fly beside the water.

The pack up complete, Margaret set off with Gari and Utu in Utu's canoe. With three people and all our gear, not to mention the rock samples I had collected, the canoe had only a few inches of freeboard. Ian and I waited on the Makalia shore, watching anxiously as the canoe made its way slowly across the open water to the bottom of the track back to Bulu Muri. Once there Gari accompanied Margaret back to the village, where he arranged for retrieval of our "*kago*" from the lake shore. He also made sure the *haus kiap* was ready for re-occupation (Ian would sleep in a small tent on the beach alongside us).

Outrigger canoes are difficult enough to paddle with two men but with one the task is even more challenging, given the constant drag of the outrigger. Nevertheless, Utu somehow managed by himself to bring the empty canoe back to collect us. With Utu paddling up front and me trying to match his pace paddling at the rear, Ian sat astride the canoe's platform as we set off to examine a part of the caldera I had not yet been to, at the far southern end of the lake's eastern arm. On the way we passed the cleft with all the bats I had shot at on my previous visit.

"Watch this," I demanded, as I lifted up the .22 and fired into the cave. Once again, the sky was quickly filled with bats, flapping over our heads and shitting on us in anger. Ian was less than impressed.

"Arsehole!"

***

On Monday, a week after arriving in Bulu Muri, the three of us stood in drizzling rain on the shore near the *haus kiap*, awaiting the arrival of the council speed boat, which was due around 10.00 am. We were pleased to be on our way back to the (relative) civilisation of Talasea, as we had completely run out of food. Margaret had carefully planned for our provisions to last the week and no more, as we had limited carrying and storage capacity and, of course, no refrigeration. When the boat had not come by midday I began to worry. Ian began to grumble.

By 3.00 pm it was obvious the boat was not coming. There was simply nothing we could do but wait till the next day. Frustrated, all three of us went through a period of fury that finally gave way to resignation and a resolve to make better allowance for the "New Guinea Factor" in future. Margaret was the most philosophical of the stranded trio:

"I have no doubt that when we finally get back to Talasea …"

"If!" interjected Ian.

"…our experience will be greeted with knowing nonchalance by the other expats."

I could already hear Harry Humphreys:

"I'm not a bit surprised," he would say, "You simply can't rely on the buggers."

Not one to hide his wrath, Ian seemed to take vicarious pleasure in reminding me of just how many transport arrangements had fizzled during his short stay with us. I tried to console him:

"Margaret heard on the radio[21] this afternoon that today is Commemoration Day, so perhaps the boat driver has taken a holiday."

Holiday or not, we were hungry and urgently needed to find something to eat. So, while Margaret went off to buy some *tapiok* or *kaukau* (sweet potato-like root vegetables) from the village women, Utu took me fishing out on the reef, where I managed to haul in a couple of good sized fish.

"At least we have survival rations," I proudly declared on returning to shore and placing my fish next to a couple of large *kaukau* Margaret had procured.

"Just what do you propose I cook with?" asked Margaret, adding, "we've got no oil or fats left for frying and no alfoil for baking the fish."

"Well I guess we'll just have to boil both *kaukau* and fish in our pots."

It was not a great success! The fish curled up and became leathery and unchewable, while the *kaukau* swelled up into a gluggy, unpalatable wad of starch that was impossible to swallow. Ian, already in a vicious mood, refused to eat any of it and went to bed hungry.

"And thoroughly pissed off!" he yelled from his tent.

***

---

[21] We had a small transistor radio with us.

Just after eight the next morning we sighted a small cargo boat heading north around the coast so I quickly had Gari and Utu paddle me out in a canoe to intercept it. On board the boat I was able to radio the Kiap in Talasea. Willhelm told me that he thought the council boat had left an hour earlier and promised to send for us if we had not appeared by that afternoon. Back on shore I related this to Ian, whose frustration was bubbling just below the surface:

"And I'm still bloody hungry!" he added, through gritted teeth.

Margaret was calm but relieved that I had been able to make contact with the Kiap. A short time later the speed boat arrived and we quickly bundled our stuff into the boat and set off. But before we did so, I quizzed the driver about why he had failed to show up, as arranged, the previous day:

"*Bilong wonem yu no kam astadei?*" (Why did you not come yesterday?)

"*O Masta, mi sori tumas. Mi gat pen long het bilong mi,*" came his quick apology.

Ian's Pidgin was minimal but he well and truly understood what had just been said, summing up the frustration we all felt with his snide reply:

"You're pain in the head gave me a pain in the arse!"

I think the boatman got the message.

Prof Ian Carmichael sits on Utu's canoe as we navigate Lake Dakataua

16

THE MASALAI

It was late September and Ian Carmichael was preparing to leave us. After returning from Bulu Muri he and I spent several days examining rock outcrops and volcanic features in the broad vicinity of Volupai and Talasea. It was a welcome opportunity to resolve geological issues that had been puzzling me since my arrival in Talasea. I also took the time to compile the information I had mapped on my set of air photographs into a geological map of the peninsula. Ian seemed very pleased with the result. For all his bombast, Ian was always quick to praise what he considered to be good work and I, still young enough to crave the approval of my superiors, gladly accepted his compliments. Margaret, too, had begun to see why I admired the man so much. He set and expected high standards and was never slow to draw my attention to any adverse outcome, but there is no doubt that he brought out the best in me.

"He seems to like keeping you a bit on edge," Margaret murmured to me in late night pillow talk.

"Yeah, it's quite nerve-wracking, but it does make me think pretty carefully about what I'm doing and saying."

Over the preceding weeks, we had listened to him lavish praise on his other graduate students at Berkeley, telling us a lot more about the man than he would ever have volunteered directly. As we delved into his scientific philosophy we could see a remarkable intellect shine through his often confronting personality. He was clearly a man wedded to his science but even more, he was devoted to excellence in that scientific pursuit. The result was a character of great complexity where intellect smothered insult. We took every opportunity we could to expose that intellect to Talasea 'Society'.

Whether in a crowd at the Talasea Club, having dinner with the doctor, Stuart Bartle, and his wife, or enjoying evening drinks with Harry and Thelma Humphreys, Ian was congeniality personified. With engaging humility, he thanked them for all the help they had given to Margaret and me, charming everyone in the process. Harry Humphreys was clearly quite chuffed at having a professor from America (albeit an Englishman) staying on his plantation and he made sure that the other residents knew just how helpful he, Harry, had been. As the beneficiaries of that generosity, Margaret and I were more than happy to let Harry hog the limelight.

On September 25th, Ian boarded a TAA DC-3 at Volupai and flew on with it to Rabaul, where I had arranged for Bob Heming to meet and host him for a visit. They apparently got on very well, for within a year Bob had accepted Ian's invitation to come to Berkeley and undertake a PhD. The Rabaul caldera would be his research topic. By the time that happened Margaret and I were well established back in Berkeley. Margaret was working at Blue Cross in Oakland, which meant that we had an adequate income. It was good to be able to help Bob and his wife, Jeanette, settle in and to repay them for all their help in PNG. Their six-year-old daughter, who had been at school in Rabaul, attended the local school in Berkeley and came home on her first day to announce:

"I've made lots of friends and my best friend is a native!"

***

But that was in the future. Meanwhile, back at Talasea, there remained several key tasks to complete before our departure. Top of the list was an ascent of the tallest volcano on the peninsula, Mt Wangore (or Mt Bola as it was known to the expats), which stood out as a prominent cone-shaped mountain, rising 1160m above the sea. None of the expat residents in the area had been up or even onto Mt Wangore and the native people also shunned it because they feared the *Masalai*.

Myths and legends abound in Melanesian culture and the Wangore *Masalai* featured strongly in the folklore of this region. '*Masalai*' basically means 'devil' or 'demon' and in this context, it was applied, despite the missionaries' objections, to a '*bikpela snek, em i laik kaikai olman*' – a giant snake that was believed to inhabit the peak, devouring anyone who dared to come near it. The myth had been reinforced by an event a couple of years earlier, as Willhelm, the Kiap, had explained to me over a beer at the Talasea Club one evening:

"A friend of mine from Holland came to Talasea looking for adventure. He saw Bola as a suitable challenge and set off one day to climb it, without bothering to hire native guides. Unfortunately, he had a fall and cut himself badly with the machete he was carrying."

For the expats, it was a simple case of under-preparation and overconfidence. But the native people saw the event as proof that there really was a *Masalai* on the mountain and attributed the injury to an attack by the monster. For me to climb the mountain successfully I would need at least two native assistants and I wondered whether I could find two such people in the face of the genuine fear that the name Wangore invoked amongst the locals.

But I had underestimated Utu. He knew all about the myth and said that, although he believed it, he was prepared to climb the mountain with me if I carried a gun all the way. Then a second guide emerged in the person of Gabriel, a strong young man from the village of Voganakai, who I had met on a previous excursion to the coast near his village. He spoke good English, which indicated a more substantial education than most of his compatriots and it was probably this that enabled him to discount the myth as exactly that – a story descended from obscurity and perpetuated by superstitious villagers. Just the same:

"I'm very happy that you will be taking the shotgun with us, Mister Garry."

Mt Wangore

EARTH

The day after Ian left us I departed Volupai in the Council speedboat. Utu, who had come down from Bulu Muri, and Gabriel, who had walked in to Volupai plantation from Voganakai, were with me, together with a couple of Voganakai mutts. The journey up the west coast of the peninsula went quickly and by midday we were setting up a fly camp on the beach below Mt Wangore. A short walk along the coast enabled me to investigate the bottom end of a large lava flow that formed a prominent ridge from top to bottom, down the western side of the mountain. As I explained to Gabriel at the time:

"This volcano is a smaller version of what once stood where Lake Dakataua is now. A great many eruptions produced lava flows like this one, building up into a giant mountain two or three times as high as Wangore."

He was intrigued.

"Where did it go Mister Garry?"

"That's a good question, Gabriel..." I was about to explain when Gabriel interrupted:

"Did it go to Vitu?"

He was referring to Garove, one of the islands in the Vitu Group, about 70km to the northwest of us. Garove comprises a caldera, breached and open to the sea on the south. The island's size, shape and orientation uncannily match those of Lake Dakataua and local legend has it that the Dakataua mountain left the peninsula and swam underwater before popping up in its present position as Garove Island.

"No," I laughed, hoping I did not sound too condescending, "that's just a myth."

Drawing pictures in the sand, I tried as simply as I could to explain the origin of calderas like Dakataua. In response, he nodded and looked interested, but I doubt he really had much idea of what I was talking about. Utu even less so. The event I was describing seemed so extreme, so unimaginable, so incredible that even I struggled to comprehend it. Like most people, well-educated or not, Gabriel saw the landscape as essentially immutable once formed. He found it hard to understand the concept of a dynamic earth where change is constant but operates on a timeframe vastly different from our own.

My conversation with Gabriel did not end there. As the three of us sat on the beach eating a basic dinner, he started to talk openly about his family, his high school education in Cape Gloucester and his hopes for the future. With Papua New Guinea heading for independence, although it was still eight years away, I wondered whether Gabriel would rise to take a leadership position in a newly independent nation.

Dusk is very brief in low latitudes such as these and as we sat there I could almost hear the sizzle as the sun sank quickly into the Bismarck Sea. For just a few minutes the western clouds shone brightly, layered in reds and yellows. The evening rays bathed all three of us in a warm glow that seemed to pick out and highlight the smooth structure of Gabriel's young face and the lines of experience etched into Utu's older visage. Their brown skin became radiantly golden in the fading sunlight and I thought of these people as a handsome race, well adapted to their environment. I knew I had the respect of these two Melanesian men, such was their colonial upbringing, but after dinner, as we sat near the fire on the beach, the barriers of race and class and social status dissolved into the darkness. We were just three mates swapping yarns around a campfire. As I listened to them I felt privileged to share a level of trust that they might never have given to their more traditional *"Mastas"*. With their bright eyes shining in the firelight, they related tale after tale, immersing me ever deeper into Melanesian culture. Stories of spirits and ancestors, of the land, the sea and the sky, of birds and trees, of fiery mountains and how things came to be the way they are. Their endeavour to understand the world around them mirrored my own fascination with earth history. Even now as I recall the experience a little shiver of pleasure tingles the back of my scalp.

The next morning it was back to reality. After a quick breakfast, we set off around 6.30 am to scale Mt Wangore, which sat looming over us, like a brooding beast of prey. Gabriel carried a day pack with food and water and I carried my geological hammer and the shotgun, as I had promised. There was no track to follow so Utu, spared the need to carry my large sledge hammer, took the lead, slashing a path through the undergrowth with his machete. The going was tough and as we made our way higher it became increasingly steep. But by then I was very fit and found the physical challenge well within my capacity. The two dogs from Gabriel's village accompanied us, yapping and panting around us as we made our ascent.

By 11.00 am I stood at the summit of Mt Wangore, on a narrow ridge between the outer slope of the mountain behind me and the floor of the crater, some 50m below, in front of me. It was a different world up there, quite unlike anything I had seen previously on the Willaumez Peninsula. The forest dripped with moisture as thick cloud drifted amongst the trees, creating weird shapes and shadows in the half light. Moss grew long and shaggy on the tree trunks, competing with fungi, some brightly coloured, some mottled brown or dirty white. Beneath my feet the leaf litter reeked of impermanence. The only sound was the panting of the dogs.

In this eerie atmosphere, I could almost believe in spirits and demons but much to Utu's relief, there was no sign of the *Masalai*.

"*Mi tingting i save masket bilong yu, na poreit longem,*" (I think he knows about the gun you have and is frightened of it) declared Utu. I saw no point in arguing with this logic as we stood shivering slightly in the chill, damp air.

"We need to keep moving to warm up," I announced, as I set about collecting some samples of the rocks exposed in the crater wall and entered my geological observations into a notebook. After a couple of hours, we had all I needed so we sat down on an outcrop of lava and ate our meagre lunch. I was relaxed; Gabriel was calm but cautious; Utu was fidgety, nervous, anxious to move on. He clearly expected the *Masalai* to jump out of the bush at any moment. I tried to reassure him, patting the shotgun that lay across my knees.

"Don't worry Utu, if the *Masalai* comes I'll shoot it." Utu did not look overly convinced, even after Gabriel translated into the local language.

Soon, happily for Utu, it was time to begin the descent. Walking down such a steep slope can be harder on the legs than climbing it, with considerable effort required to avoid tripping and falling. The dogs once again went ahead of us, four legs obviously better than two in that environment. About an hour into the descent the barking and yapping from the dogs increased in intensity. We could not see them but they apparently had found a wild animal that they were harassing in the undergrowth nearby.

"It could be a pig," suggested Gabriel, "can you please shoot it if the dogs chase it to us?"

"Of course," I replied, happy that the shotgun I had laboriously carried all the way up the mountain might finally be used to advantage. Excitedly I loaded a cartridge into the barrel and closed the breech, ready for action. Whatever it was it was coming closer, as the dogs continued their noisy chase. Suddenly the bushes near me parted and a cassowary sprang out in front of the pursuing dogs.

"*Siutim Masta!*" yelled Utu.

Immediately I aimed at the cassowary and pulled the trigger, flinching for the bang. Nothing happened. I pulled the trigger again. Same result. The cassowary, seeing us, did not hang around. Within seconds it dived back into the forest, still pursued by the dogs but gaining extra adrenalin for escape after encountering the three humans. A few moments later the barking subsided and the dogs returned to us, panting and excited but without a catch. I examined the shotgun, trying to work out why it had not fired. As I did so there was a click. It appeared that I had not quite closed the breech properly after loading the cartridge and so the gun would not fire. My moment of hunting glory was gone and I felt humiliated in front of my Melanesian friends, who would have relished the feast that a cassowary could have provided, to say nothing of its colourful feathers. They made no negative comments but the silence that prevailed for the balance of the descent told me they were very disappointed at my failure to shoot the bird. Back in the camp that evening the conversation was subdued and the euphoria I had felt the previous night seemed like a distant memory.

At 9.00 the next morning the Council boat came to retrieve us and we made the journey home rapidly and in relative silence. At least I did manage to catch a couple of good sized fish on the way back, giving them to my helpers as some form of consolation.

A week or so later I ran into Gabriel once again.

"Utu thinks it was the *Masalai* that stopped you from shooting the cassowary," he told me.

Perhaps Utu was right. Does a mythical spirit have to have a physical presence to spook you?

My companions for the ascent of Wangore

Utu

Gabriel

17

BULU DAVA

By early October my field work was almost done. A large suite of rock samples had by then been collected and we began the process of packing them up for shipment to California, where I would use them as the basis of my continuing research into the volcanic geology of Talasea. The plan was for Margaret to go to Rabaul to arrange the shipping while I made a brief visit to another volcano on the PNG mainland – Mt Lamington, of which more later. We would then meet up in Port Moresby. Before all of that could happen, one major task remained.

The Bulu people had settled in two villages at the northern end of the Willaumez Peninsula. We had already spent considerable time on the east coast at Bulu Muri; now we were to go to Bulu Dava, on the coast to the west of the caldera, to map and sample the geology of that area. Once again, our mode of travel was the Local Government Council speedboat, which was brought across from Talasea to the west coast at Volupai by tractor and trailer. By 9.00 am on Monday, October 2nd, we were on our way.

Two hours later we arrived at Bulu Dava and were given a boisterous welcome by the villagers. It appeared that our experience in Bulu Muri had filtered through to their cousins in the west. Not only did we receive a lively welcome but we were very pleased at the contrast between this village and that in the east. Their Councillor was an impressive fellow, who had already arranged guides and carriers for us and had ensured the *haus kiap* was ready for occupation. Compared with the Bulu Muri house, this one was a mansion. It was much bigger, with a respectable shower house adjacent – tall enough to hang the shower bag properly – and a well constructed pit toilet nearby. Inside the house we were thrilled to find a strong table already in place and a huge bed, measuring 6ft by 7ft.

"It looks strong enough to dance on," was Margaret's immediate reaction.

"Well if we do we'd better not fall off," was my response. The bed stood three feet off the floor.

Our delight at being in Bulu Dava was enhanced further late in that first day by the arrival of Gari, from the sister village, who had come around the coast in his canoe, with his wife, to stay while we were there. He wanted to help me in the field work and offered his canoe for transportation purposes. Margaret and I were really pleased to see him and he seemed flattered by the enthusiasm of our response. We in turn were flattered by Gari's thoughtfulness and consideration of our needs, which belied the oft-repeated mantra of the Talasea expats, whose jaundiced view of the native people could be summarised as:

"They're a useless lot. You have to do all their thinking for them."

Our experience with the local Melanesian people was just the opposite most of the time (not quite all the time I have to admit; Peter for one was pretty thick). I believe now it was our deep immersion into their way of life that made the difference. For most of the expats it was "them" and "us" in the traditional colonial style. For Margaret and me, it was just "us", as we shared their living space, their food and their culture and were the happier for it.

As if to emphasise that immersion, on our fourth night in Bulu Dava we were treated to an impromptu concert. Around 7.00 pm a group had gathered on the beach in front of the *haus kiap*. With our Pidgin by then pretty good, we could engage in a substantial discussion about their way of life. Topics included children, education, crops and soil, other villages and the local diet, which included the best pineapples I have ever eaten, before or since – small, fragrant, deep yellow fruit, as sweet and juicy as a ripe peach. As we chatted the crowd grew and the children became restless and noisy. Margaret asked for the kids to repeat the singing they had done for her at the village school two days before, when I was in the bush.

"*Mipela laikim ol pikinini mekim singsing olesem hapastadei*," (We would like all the children to sing for us the same as they did the day before yesterday) Margaret requested.

It took a bit of encouragement to get them started but once they got warmed up we could not have stopped them if we had wanted to. It was quite a surreal experience hearing familiar songs sung in the local dialect, in Pidgin or even in Melanesian-accented English. They started with *Waltzing Matilda* in our honour and continued with such varied fare as *You are my Sunshine, In New Britain, Mat-til-da, Onward Christian Soldiers* and *These Little Hands*. As they sang, their parents looked on, beaming as brightly and proudly as any suburban Sydney Mum and Dad would do at a primary school Christmas concert.

On Saturday the speedboat was back to pick us up for the return journey. Unfortunately, the sea was rough that day and we were totally soaked and salt-encrusted by the time we reached Volupai. Regardless, as usual, I trailed my 250lb breaking strain fishing line behind the boat. Along the way, the boatman's hat blew off and as he circled round to retrieve it my fishing line became hopelessly ravelled and tangled. It would never be useable again, but hauling it in the one last time I was excited to find a tuna securely hooked onto my trusty white lure.

This was the same lure that had served a very different purpose, just a few days beforehand, when it became the means by which our, by then, well developed survival skills could be demonstrated.

WATER

It was near bedtime after a physical day hiking through the rain forest. Margaret picked up our torch and climbed down from the hut to visit the toilet, as she had done before retiring on preceding nights. That evening, however, I was startled by a sudden scream that emanated from behind the palm fronds.

"What's up?" the solicitous husband asked.

"Something just flew into my face," she called, "and the seat is crawling with ants. They're all over me."

"Well, be sure to brush 'em off before you come back up here," I requested, the darkness hiding the smirk on my face.

"There's a problem," came her further call, "you'd better come and see for yourself."

Crouching low, I joined her inside the little *haus pek pek*.

"So what's the problem?"

"I dropped the torch .... into the toilet."

"Oh shit!"

It was our only torch and we could not afford to lose it. Fortunately, it was a waterproof torch, encased in rubber. As I peered down into the depths of the pit I could see that it was still on, as an eerie khaki light shone in the murky water at the bottom, two metres below us.

"Hmm! We have to get that torch back Margaret, or we will be in the dark for the rest of this trip."

"And just how do you propose to do that?" her voice cracking a little as the tension rose.

"I have an idea," I said, after thinking for some minutes about what I could use to try to pull the torch up out of the water.

Returning to the *haus kiap* I found the fishing line I had been using to troll from the boat as we had travelled up the coast. It had a large white lure with two sharp hooks attached. Back in the *haus pek pek* I lowered the lure into the murky water and began jiggling it around. At first all that happened was movement of the light, as I nudged the torch but did not hook onto it. Then,

"Strike!"

I had hooked onto the rubber casing and began to pull the line in. Our hopes rose in tandem with the torch as it emerged from the water. Then,

"Splash!"

The torch dropped off the hook and fell back into the water.

Suddenly Margaret was hearing words very new to her.

Oblivious to the smell and the ants that were crawling up my arms, I peered into the depths once more, dunking and wriggling the lure and eventually hooking up a second time.

Slowly the torch rose out of the water, swinging around and around, its light making creepy images on the walls of the pit. Hardly daring to breathe, lest I shake the torch off the hook again, I carefully retrieved the line. As it came nearer I was tempted to reach down and grab it but Margaret wisely held me back. Come to think of it, she has done that quite a few times over the years, to my great benefit.

Then I had it. The smelly, slimy thing was in my hand, its light still shining brightly, and our crisis was over. Breathing again, reluctantly, as the air was pretty ripe, I turned and looked at Margaret.

"That was fun, wasn't it?"

Grimly, she repeated some of the words she had just learnt from me.

Relieved, in more ways than one, I switched the torch off as we exited the *haus pek pek*. Then it was down to the shore to wash the fetid thing in sea water. After that Margaret made up a dish of Dettol solution and sterilised the torch by soaking it in the solution overnight.

That torch continued to serve us well for the rest of our time in PNG and for years afterwards. And the fishing lure, returned to its rightful purpose, had one more moment of glory to come. When it did so, I pondered whether it carried a residual aroma that made it more alluring to the fish.

Bulu Dava school kids and their teacher

18

SLEEPING GIANTS

What constitutes an active volcano? How can you tell that a volcano is extinct, rather than merely dormant? Russian volcanologists consider any volcano that has erupted within the last three thousand years as 'active', while the Japanese extend the definition of 'active' to the last ten thousand years. Such timeframes are well beyond any human lifetime; they also push back to pre-history, where our knowledge of human existence and the natural disasters it suffered is fragmentary at best. Most people base their judgements and reactions to natural phenomena on the experience of their own lifetime, or a few generations at most. They see the earth as static and stable, as old and unchangeable as the hills. That is understandable, but it is a poor metaphor, as the hills are far from unchangeable. Although the Blue Mountains in New South Wales may look the same today as they looked 70 years ago, they are not. And they certainly are not the same as they were 700 years ago, or 7,000 years ago or 7,000,000 years ago. Perhaps the greatest enlightenment that a study of geology bestows upon a person is an appreciation of the nature of time, deep time, geological time – the fourth dimension of Albert Einstein that I referred to in Chapter 1.

Enlightened so, one can see catastrophic events such as the 2004 Boxing Day tsunami (caused by a major subduction earthquake near Sumatra) or the Fukushima subduction earthquake and resulting tsunami in March, 2011, less as sudden cataclysms and more as ongoing natural processes that have shaped, are shaping and will continue to shape our dynamic world. This is as true for the deep-seated forces that drive volcanic activity as it is for the tectonic calamities that we see as earthquakes. Both are inevitable consequences of convection currents in the earth's mantle that flow up from the core at about 6 cm per year. That may not sound like much (it is about the same rate as your fingernails grow), but it is enough to crush continents together, or split them apart, as they bring the heat of the core to the base of the crust and form the magma that we see as lava flows.

Are those volcanoes at Undara in North Queensland extinct, or have they just been dormant for the last 12,000 years? Is Tower Hill in western Victoria safe, now that 4,000 years have elapsed since the last eruption there? Even where human history is well documented, eruptions can come as a surprise, with devastating effects – witness the unexpected eruption of Mt Vesuvius[22] in 79 A.D. and the resulting destruction of Pompeii and Herculaneum. The account of this eruption by 'Pliny the Younger' (*Gaius Plinius Caecilius Secundus*), in which he describes the first-hand investigation of the event and consequent death of his uncle, 'Pliny the Elder' (*Gaius Plinius Secundus*), makes fascinating reading. The account concludes:

> *"They thought proper to go farther down upon the shore to see if they might safely put out to sea, but found the waves still running extremely high, and boisterous. There my uncle, laying himself down upon a sail cloth, which was spread for him, called twice for some cold water, which he drank, when immediately the flames, preceded by a strong whiff of sulphur, dispersed the rest of the party, and obliged him to rise. He raised himself up with the assistance of two of his servants, and instantly fell down dead; suffocated, as I conjecture, by some gross and noxious vapour."*

---

[22] The Romans had no idea that Vesuvius was a volcano, as it had not erupted within their historical memory.

In a country like Papua New Guinea, where documented history is very brief, the potential for surprise is even greater. With so many volcanoes, including some that were not recognised as such until too late, unexpected eruptions should be expected. My own experience during those few months in PNG in 1967 exposed me to two excellent examples. One was within my study area at Talasea, the other was on the PNG mainland. At that time, the first event lay in the future while the second was in the recent past.

FIRE

On 16th October, 2005, nearly 40 years after I was there, a strong earth tremor was felt at Talasea. It was not a great concern. Earth tremors, or *"gurias"* as they are known locally, are common enough in this tectonically active region. The following day though, just 16km south of Talasea, Mt Garbuna suddenly erupted, sending a column of ash 4km into the atmosphere. The mountain was well known as a volcano. Indeed, my own footprints were embedded in the soft hot ground at the top of the mountain, where the largest geothermal field in PNG was and still is very active. Such a large and vigorous geothermal field implies the presence of a substantial heat source at shallow depth but in 1967, there was no sign of recent eruptions. Since my time there, other geologists have deduced that in 2005, Garbuna had been dormant for nearly 1,800 years. The mountain's re-awakening after that time served to underline the brevity of human experience and the longevity of geological phenomena. Eruptions continued at Garbuna through into November of 2005 and then subsided. In March, 2008, activity resumed on a smaller scale and lasted until September. Today it is again quiet, but for how long?

Mt Garbuna, with its southern neighbour Krummel and its northern neighbour Welcker, is the middle one of three volcanic peaks that together constitute a large shield volcano. In late August, 1967, I set out on a three-day exercise to climb Garbuna and inspect the geothermal field that was known but rarely visited and never studied. At that time, there were no historical records of eruption, nor any seismic activity that might warn of eruptive potential. With Martin from Volupai and two other assistants from the village of Kilu, on the east coast below the mountain, I began the ascent. The first part was easy, walking through native gardens and copra plantations to reach the edge of the rain forest. The villagers working in their gardens looked up in surprise but greeted us warmly as we passed by in the airless morning heat. They were curious about where we were going but showed no interest in tagging along.

With the gardens behind us and the ground sloping steeply upwards, it was then a case of hack and cut through the undergrowth. It was an arduous but by then familiar experience for me, as our little band made its way up through the rain forest.

After a couple of hour's effort, we reached the open air at the top, where the thermal activity had prevented any vegetation growth. As we walked into an old crater all our senses were assailed at once. First the smell; rotten egg gas[23] wafted strongly on the breeze, making us recoil in distaste. Then the noise; all around us steam was escaping from vents in the ground, roaring and hissing as though the ground was alive and resented our intrusion. After that the heat; billowing clouds of malodorous condensed steam enveloped us, warm, damp and alien. And finally, the visual spectacle; looking across the crater revealed countless bubbling, boiling pools and mud pots, interspersed with vents ejecting steam

---

[23] Hydrogen sulphide – H_2S.

under high pressure. It was easy to see why the term "hell" is so often applied to such geothermal fields; for me, it was heaven.

At the bottom of the crater a small creek flowed with steaming hot water that was painful to walk through, especially for my assistants in bare feet. The rocks and soil within the crater were altered, bleached and stained by the thermal waters, yellow, brown, red and white. Most striking of all was the sulphur, which stood in mounds around fuming vents. Bright yellow, its crystals glistening in the sunlight, some of the sulphur mounds were as tall as a man. My native assistants had never seen anything like it. Nor had I. They were fascinated and full of chatter as late in the day we retired to our small camp in the forest just outside the crater.

The next day we trekked over to an adjoining crater, where the scene if anything was more extreme than the one we had seen on the first day. Adding to the extra-terrestrial atmosphere this day was a change in the weather. Low cloud drifted across the crater valley, at times masking and at times revealing the activity. Then it began to rain. Soaking rain was nothing unusual for me by then, but it did make the conditions more bizarre, as the damp air and cool water falling heavily from the sky mixed with the hot vapours and boiling water issuing vigorously from the ground. The roar from above competed with the roar from below as we trudged between the two. My enthusiasm for the spectacle around me began to dissolve in the water dribbling down my face. We stuck it out for as long as we could but then I decided that it would be wise to retreat to our little camp. Not that a small tent fly was much of a shelter. When I pulled out my thin blanket to provide some warmth I was disgusted to find that it had been fly-blown in our absence, with a dozen or so soft white eggs embedded in the fabric. Life in the rainforest misses no opportunity for reproduction. We spent a rather miserable night, made all the more unpleasant by the fact that I had forgotten to bring an opener for the canned food we had with us. Fortunately, the point on my geological hammer ("G-pick") was sharp enough to penetrate the lids, but it was less than ideal.

The following day we returned wet and bedraggled to Kilu village, where I was able to get a lift for Martin and me back to Volupai. Margaret welcomed me home with a delicious hot dinner and a freshly baked cake. Martin was happy to walk back to his village, grinning widely as he carried a large portion of the cake.

***

About twenty years later I was back in New Britain as an exploration geologist, investigating some mineral prospects that had been offered to my company. On that occasion I was travelling by helicopter and, being already near Talasea, I decided to make a brief side trip to revisit some of my old haunts. While working as a mineral explorer paid the bills, and I enjoyed it, my infatuation with volcanoes remained unabated. Top of the list was a look at Mt Garbuna. We flew low over the scarred landscape, with vents and pools steaming as much as ever. From the air, it was possible to see just how extensive the thermal activity was. Even from above, the sulphur mounds glistened in the sun.

"Can you put us down over there?" I asked the pilot, pointing to a relatively quiet, slightly elevated spot within the crater.

"Ahhh! I dunno. Is it safe?" he asked.

"It'll be fine," I replied, "I've walked on that ground and it's quite solid enough."

"It bloody better be!"

The pilot set the chopper down, gingerly, as though expecting the ground to give way, or to explode beneath him. As he shut the engine down my colleague, Terry, and I climbed out of the helicopter and began to look around in wonder. The pilot remained close to his chopper, ready for a quick getaway, with or without us. Terry and I then spent an hour or so inspecting and photographing the exotic features, most notably the sulphur columns.

Little did we realise that in another twenty years or so — a mere blink of a geological eye — the very ground we stood on would be rocked by earthquakes and blasted into the sky as the volcano I thought was extinct burst back into life.

Sulphur and steam inside the Mt Garbuna crater

FIRE

It was Sunday morning, 21st January 1951[24]. All over the Popondetta district the Orokaivan people were attending church. This part of Papua New Guinea was well settled as its fertile soils were renowned for agricultural productivity and there were many villages, some with substantial schools where children boarded as they undertook their education. The Australian Administration, under District Commissioner Cecil Cowley, was in control as he and his staff guided the Orokaivans and brought the benefits of post-war development to an area so recently ravaged by conflict and the Japanese occupation. Popondetta was, after all, the northern gateway to the Kokoda Trail.

Looming over the district was a mountain known to the Orokaivan people as *Sumbiripa Kanekari* – the "Separation of Sumbiripa." According to legend, it was where Sumbiripa became separated from his wife, Suja, while hunting. They found themselves on different parts of the mountain where they unfortunately died, in the process imparting their facial features to their respective parts of the mountain. In doing so, Sumbiripa became the Master of the Mountain. To the Orokaivans, this mountain held a focal position in their world order. For them, it was the centre of the cosmos and the place where, according to their myths, death, warfare and fire originated.

To Mr Cowley and the other European settlers, teachers, missionaries and government employees it was simply Mt Lamington, an isolated peak about 5,500ft (1,680m) above sea level, not unusual or remarkable in this land of mountain ranges rising to nearly 15,000ft. True, it did sit out by itself to the north of the Owen Stanley Ranges

---

[24] This account draws on an article by McLaren Hiari, published in the *Sunday Chronicle*, Papua New Guinea's weekly newspaper, on 23 January, 2010.

and a little to the west of the Hydrographers Range but jungle-clad mountains like this are legion in Papua New Guinea. Mt Lamington was nothing special, nothing threatening.

A few days beforehand, on 15 January, some local people had reported seeing smoke and glowing rocks being ejected from the top of the mountain, forcing the government officials to acknowledge that Mt Lamington was in fact a volcano. But no warnings were issued, no volcanologists were called in and no evacuation advice was forthcoming. As the mild eruption of ash continued, accompanied by earth tremors and landslides, Commissioner Cowley issued a reassuring statement:

"There is no loss of life nor is there any immediate danger."

Both the native and expatriate communities thought the ongoing activity at the top of the mountain was evidence that pressure was being released gradually, reducing the likelihood of a major eruption, even though the intensity of the eruptive activity was slowly increasing. On that fatal Sunday morning messengers were sent to the churches carrying notes to be read out during the services, offering reassurance to the villagers that there was no need to worry. By 10.30 am the church services were in full swing and the Melanesian proclivity for hearty singing rang out over many villages and boarding schools.

At 10.40 am their world changed forever. A giant, paroxysmal explosion rocked the nearby mountain and the entire district. Its sound was heard 100km away. Survivors described the column of ash as being "seven miles high" as the northern side of Mt Lamington was blasted into the sky. But worse was to come. The column of ash began to collapse under gravity, forming a huge pyroclastic flow[25] that swept down the mountainside and spread across the district for

---

[25] A hot dense cloud of volcanic gas, ash, dust and rocks that rolls downhill at great speed.

up to 15km. For 3,500 people, including 35 Europeans and over 200 school children, death was instantaneous. Twenty nine villages were annihilated. The Higaturu Government Station, the Commonwealth Rehabilitation and Training School, Sangara Anglican Mission Station, Martyrs' Memorial School and Sombou Primary School all simply ceased to exist. So did District Commissioner Cowley. Later, victims were found with their clothes stripped off by the force of the ash flow. At Higaturu a jeep was thrown up and wedged between remaining branches of a skeletal tree. The devastation was complete and covered an area of 230km^2.

In size and force, the Mt Lamington eruption of 1951 was comparable with that well recorded more modern example at Mt St Helens, in Washington State, USA, in 1980. A detailed scientific analysis and report on the Mt Lamington eruption was written by an Australian Government geologist called Tony Taylor, who was on the scene soon after the main explosion. Taylor documented the event and its aftermath in great detail[26], even as the mountain continued to erupt. He was awarded the George Cross for his bravery in investigating the Lamington eruption.

Activity continued at Mt Lamington after that cataclysmic event but gradually declined until it ceased in 1956. Eleven years later I stood atop this mountain, marvelling at the scene in the crater before me.

It was curiosity and that enigmatic human trait – an innate desire to climb upwards – that had drawn me there. I had wanted to see Mt Lamington ever since my aviator brother, Graham, who had often flown by the recently active volcano when he was in PNG, had told me about it. Ian Carmichael agreed that I should visit the mountain before leaving the country and the arrangements to do so had been

---

[26] *"The 1951 Eruption of Mt Lamington, Papua"* by G.A. Taylor, G.C., published in 1958 by The Bureau of Mineral Resources, Geology and Geophysics as Bulletin 38.

made with assistance from Bob Heming.

"You need to go to Popondetta and then get yourself to the village of Kendata," he had explained before I left Rabaul.

"Contact this bloke when you get to Popondetta and he'll arrange a car for you," he added, handing me a slip of paper with a name and phone number.

"You'll be able to camp overnight in the *haus kiap* at Kendata. Start early as you can the next morning because it's a 7-hour climb from there."

Others had made the climb before me and although there was no track as such, I was assured that local guides in Kendata would be able to show me the best way to make it to the top of the volcano.

Travelling to Popondetta involved another hair-raising flight in a DC-3, this time north from Port Moresby and across the Owen Stanley Range. I remember looking out the window of the aircraft as it struggled over a high pass. Through breaks in the cloud I could see trees just off the starboard wingtip; out to port the view was the same. Looking down the trees looked just as close. Something more than concern but not quite terror ran through my veins.

"I'm sure we'll make it," I kept telling myself, "they've done this many times before." And, of course, we did. The reliable old DC-3 was more than up to the task.

I do not really remember much about the actual climb. By then I was well used to cutting a path through rain forest and conversant with the hazards and challenges it brought. My Kendata guides did most of the physical cutting anyway; I just had to look after myself. The weather was kind to us and there was no rain, which allowed rapid progress up the mountain. In the end, I reached the top in five and a half hours, with only one rest stop along the way on a particularly steep pinch – my Talasea-induced fitness came well and truly to the fore. Even the native guides had trouble keeping up with me.

At the top of the mountain distant views were obscured by drifting cloud but the scene in front of me captured my full attention anyway. Vibrant forest regrowth had reclaimed the outer volcanic slopes up to the crater rim but inside it was a different story. Rising prominently in the middle of the crater was a large dome of light-coloured volcanic rock[27], which almost filled the space available. Grass and moss dotted the dome but there were no trees. Columns of condensed steam mingled with the drifting cloud, creating that other-worldly appearance that only the inside of a recently active volcanic crater can offer. I scrambled down the face of the crater wall and up to the volcanic dome. The rocks were hot to touch and I concluded that this dome had been extruded as a viscous plug after the cataclysmic eruption of 1951. It now sat as a cap on the volcanic conduit, perhaps one day soon to yield to pressure from below and erupt again in a violent blast like that which devastated the area just sixteen years before.

Today the volcano is regularly monitored by volcanologists from Rabaul or Port Moresby, who might be able to predict an impending eruption and arrange for the evacuation of people in danger. Meanwhile, Mt Lamington, or *Sumbiripa Kanekari*, sits quiescent, overlooking Popondetta, biding its time, waiting for human complacency to set in once again.

---

[27] Dacite, a viscous lava fairly high in silica.

This is what is left of Mt Lamington after the 1951 eruption
- a steaming new lava dome sits inside the crater

19

"TAIM BILONG GO PINIS"

AETHER

Dark clouds loomed on the horizon and each day the humidity seemed to increase. Soon the bright sunny days and fresh breezes of the dry would give way to the torpid air and steady rain of the approaching wet season. September had rolled into October and at the Talasea Club the mood of the expatriate community matched the lowering sky. It was time for us to leave ("go finish" or "*go pinis*").

My work was just about done and we began to say our good-byes. Richard had returned from furlough and reclaimed his house, so we moved across to stay with the Humphreys for the final few days at Volupai. Sadly, we had to say farewell to Sung-in:

"*Tenkyu Sung-in. Yu gutpela tru. Mi sori tumas, tasol mipela mas go pinis. Lukim yu bihain.*" (Thanks Sung-in. You are a good bloke. I am very sorry, but we must leave. See you later.)

His eyes glistened as I spoke, for although it was very much a master-servant relationship, as required by the times and circumstances, we had built a strong bond with this ever loyal and (relatively) hard working man. The missing cakes of soap seemed very trivial now and his scones had improved out of sight.

The last rock samples had been collected and each one carefully wrapped and packed into a crate – 349 specimens in total. When all was ready we summonsed the Council tractor and trailer for one last time to cart the crate and its hard-won contents to the wharf in Talasea, where it would be taken by coastal vessel to Rabaul for forwarding as sea freight to San Francisco. From there it would be a short trip across the Bay to the Earth Sciences Building on the Berkeley campus. It was not exactly a strong crate and would require wrapping with steel bands once in Rabaul. At the wharf, I watched anxiously as the crate, which weighed something like 350kg, was man-handled onto the boat. I held my breath as four or five men struggled to contain the weight and I feared the product of my labours over the past three months would end up on the bottom of Talasea Harbour. But all was well and I relaxed as the crew tied the box securely onto the foredeck of the small ship.

On October 12th Margaret and I parted company, as I headed south to Lae and then Port Moresby to make the side trip to Mt Lamington. Margaret headed north to Rabaul, where she was to return the borrowed gear to Bob Heming and arrange for the rest of our personal goods to be sent back to Australia. Above all, she also was charged with organising the shipment of my precious rock samples to America. That turned out to be more difficult than we had expected, as, in our naivety, we did not realise that we needed an export permit to send the rocks away from PNG. Until she had that, the freight forwarder would take no action. The required permit had to come from the Mining Warden in Wau (on the PNG mainland) and it took several frantic cables back and forth to obtain the necessary document. Somehow or other, she managed to achieve this while I, blissfully unaware of the problem, was climbing Mt Lamington.

Several days later we were reunited at Port Moresby airport. Our Papua New Guinea adventure was over. As we climbed the stairs to board the TAA flight back to Sydney I took one long and pensive last look at PNG. Both Margaret and I had developed a strong attachment to this exotic land and its friendly people.

We had survived, we had succeeded, we had made new friends and immersed ourselves in a culture stranger than anything we might have imagined just a few months earlier. The people, the places and the experiences of Talasea had enriched our lives beyond measure and forged a relationship between the two of us that would carry us through the trials and tribulations of the next fifty years.

Part of us will always be in Talasea: Looking out to sea from the *haus kiap* in Bulu Muri; canoeing over the reef off Voganakai; listening to the children sing in Bulu Dava; sniffing the rotten egg gas at Mt Makalia; coating our legs in Johnson's Baby Oil; trudging through a tropical downpour; dodging *mong mongs* in the rain forest; watching the fireflies flash at Volupai; laughing with Gari at his fear of pythons; hearing the gentle hiss as yet another stubby of South Pacific Lager is opened at the Talasea Club.

We would never see Utu or Gari or Sung-in or Gabriel again but we would never forget them. Margaret and I remain forever grateful to have shared this rare and unrepeatable experience with them.

We had already booked a passage to San Francisco on the S.S. "*Oriana*", leaving Sydney in November. There was much to look forward to, especially for Margaret, who was yet to experience the delights of California in the Sixties. But as the 727 climbed away from Port Moresby I felt quite wistful, my excitement about returning to California with my new wife tempered by sadness to be leaving this mysterious, at times frustrating, but always sublimely beautiful country.

A few hours later we were back in Sydney. It was October 16th, two days before Margaret's 22nd birthday.

My precious cargo of crated rock samples sits on Talasea's wharf, awaiting shipment to Rabaul and on to California

3. EQUATORIAL EMERALDS

With some new friends in Kalimantan, Indonesia, ca. 1983

20

SIX THOUSAND ISLANDS

AETHER

Well, actually, there are 13,667 of them altogether, but only about 6,000 are inhabited. Still, that is more than any other country in the world and makes Indonesia the world's largest archipelago. This vast chain of islands extends for more than 5,000km east to west and almost 2,000km north to south, straddling the equator and dividing the Indian Ocean from the Pacific Ocean.

The majority of the population lives on five major islands (Sumatra, Java, Kalimantan, Sulawesi and Papua) and most of the rest live on thirty smaller islands, including that favourite Aussie holiday destination, Bali. More than 260,000,000 people are spread across the archipelago, making Indonesia the fourth most populous country in the world. Some 88% of Indonesians identify as Muslim which means that Australia's nearest neighbour (except for Papua New Guinea) is also the largest Muslim country on earth. This fact was brought home to us all too clearly by several tragic events that have taken place there, especially in Bali, in the early 21st century. Even so, most Australians remain largely ignorant about this country and for many their only exposure to Indonesian culture has been in the bars and hotels of Kuta and Nusa Dua.

That is a shame, because Indonesia is a beautiful and culturally diverse country, with a rich heritage of mythology that informs its music and dance. For the most part, the land is lush and green, with abundant rainfall and fertile soils derived from the 300 volcanoes that dot the smaller islands and form the backbone of Java and Sumatra. Its geologically young and active environment has delivered a wealth of mineral resources, including gold, copper, nickel and other minerals. The people, by and large, are friendly and welcoming, especially in the villages and towns of the outlying provinces. Indonesia truly is a string of emeralds on the equator.

One way or another, this fascinating but often challenging country has played a key role in my professional career, beginning in 1973 and continuing on and off for the next 37 years, with around 20 visits in all. Initially, during the 1970s, I was there as a geologist for Kennecott Copper Corporation. Kennecott had a major copper-gold exploration program in Sulawesi Utara, the northern arm of this strangely shaped island that used to be known as Celebes. After that I spent time looking for gold in Kalimantan and then much later still, I returned to Indonesia as a director of Straits Resources, an Australian mining company with coal and gold mines in Kalimantan. After Australia and the United States, I have spent more time in Indonesia than any other country on earth and my life has been greatly enriched by my experiences there. Most of those experiences were good, some were scary, a few were unforgettable and some were quite regrettable, especially the dengue fever and malaria that Indonesia unkindly gave me.

These days Indonesia is a modern country with a highly literate population and well established transport and communications infrastructure. Forty years ago, things were a little different.

21

GETTING THERE IS HALF THE FUN

It was not my first visit to Indonesia, or indeed my last, and on this occasion, I had travelled to Gorontalo as usual by flying *Bouraq Indonesia Airlines* from Jakarta. The trip took virtually all day, leaving Jakarta at 5.00 am and arriving 11 hours later in Gorontalo. The *Bouraq* aircraft were old, cramped turboprops, either an HS-748 (a British build) or a YS-11A (a Japanese copy). They also had a Fokker F-27 for a while and there seemed to be a total of eight to ten aircraft in the fleet. That number steadily dwindled as *Bouraq's* aircraft crashed with disturbing regularity in remote parts of the country, eventually forcing the airline to fold.

The first leg of each trip was direct to Banjamasin on the south coast of Kalimantan (Borneo). Within minutes of leaving Jakarta the top half of the cabin was a haze of smoke from the free cigarettes handed out by the cabin crew. It is hard to hold your breath for three and a half hours, so I just kept my head down and concentrated on the cold rice and chicken breakfast that was served to me. The chicken was tough and chewy – the term "chicken" was used rather loosely in the Indonesian cuisine of those days. The rice was more or less edible, the better so once I had picked out the tiny black bodies of the weevils that were included as garnish.

"Just eat them," I was told, "think of it as extra protein."

"Bugger that."

The *Bouraq* breakfasts may have had something to do with the severe gastroenteritis I suffered almost every time I visited Indonesia in the 1970s. It usually came on within the first few days of arriving in the country, afflicting me with gut-wrenching, "leave me alone and let me die" type gripes for a few days but then did not return for the remainder of the visit. The only time I did not get sick was once when I returned to Indonesia after only two weeks at home in Sydney. My immune system was clearly on the job that time!

The next leg of the journey to Gorontalo was up the east coast of Kalimantan to Balikpapan, a big town with a thriving oil industry. The oil men all flew in *Garuda* jets direct from Jakarta but we unfortunate copper exploration geologists were relegated to *Bouraq*, very much the poor relation to Indonesia's national airline. Sometimes there was a substantial wait and a change of aircraft in Balikpapan. This required the continuing passengers to bide their time in a small concrete sweat box whose use in conflict would be considered a war crime but was offered by *Bouraq* to its passengers as a "Terminal". Inside it was hot, humid and smelled of human sweat and urine; outside the "Terminal" it was hot, humid and smelled of human excrement, with a bit of aviation fuel thrown in for good measure. Outside was blisteringly sunny, but at least there was some breeze. Inside there was shade but the air was stifling and static, except for the constant buzz of mosquitoes. In the vanguard of the insect attack was *Aedes Aegypti*, carrier of Dengue Fever, a virulent tropical disease that on one occasion struck me down at home a week[28] after returning from Sulawesi. I have no doubt that my Dengue Fever was a take-away gift courtesy of *Bouraq* and the long layover I had endured in its Balikpapan "Terminal" on the way home.

---

[28] Its normal incubation period.

After an hour or two in Balikpapan it was a relief to re-board the aircraft, although with no air conditioning until the engines were operating the plane initially felt like a tubular version of the boxy terminal. When the engines were started, there was always a man holding a huge trolley-mounted fire extinguisher right beside the aircraft. I was never quite sure whether to be pleased or concerned about this. I suppose it was a worthwhile safety measure but I did not find it all that reassuring. As soon as the engines were turning the air conditioning came on, causing a steady stream of condensation to dribble from the ducts in the cabin ceiling. Finally, despite all indications to the contrary, the journey continued to Palu, in the north of Sulawesi.

The most remarkable thing about Palu was the contrast between the jungle-clad slopes visible below on approach to the airport and the dry, barren grassland of the Palu valley that emerged once a sharp ridge was crossed just before landing. It was the most dramatic rain-shadow effect I have ever seen. Challenging that, however, was the disconcerting sight of the wreckage of a *Bouraq* aircraft that lay in the scrub, just beyond the end of the runway.

Then at last it was on to Gorontalo, in the northeast of Sulawesi. Or at least it was usually so but on my very first visit the rain in Gorontalo was so heavy we could not land (just as well, see Palu above). That flight had to continue to its final destination, Manado, at the far eastern tip of Sulawesi Utara, and return the next morning. On that occasion, the airline put its Gorontalo passengers up in a dodgy little hotel in Manado. As I dozed at 5.00 am the next morning, after a night of swatting mosquitoes and tossing restlessly in the humid, airless room, I was woken by a sudden blast of noise seemingly right beside me. It was the call to prayer issuing from loud speakers in the adjacent mosque – a startling introduction to Indonesian culture for a first-time visitor!

Even after arriving at Gorontalo airport it was a long drive to the city of some 60km through rural communities and small villages with countless children, chickens and mangy dogs running amok, each equally at peril from passing traffic. So, after a couple of days in Jakarta arranging a "Police Permit" to go to Sulawesi, a very long day getting there with *Bouraq* and a tedious drive in from the airport, I would finally arrive at the house in Gorontalo that was Kennecott's operating base for the Sulawesi copper exploration program.

The house we had was large, built during the Dutch colonial period, with big rooms and high ceilings over terrazzo floors. There was no air conditioning but the house was designed to catch whatever breeze was blowing so it was reasonably comfortable. There were kitchen staff to prepare meals (and, I suspect, incubate gastroenteritis flora) and people to clean, launder and carry out any other required domestic duty. There was a bathroom and a 'Western' toilet, as well as a more typically Indonesian 'squatter'. My confrontations with hostile microbes ensured I was well acquainted with both options. I remember one particularly severe attack when I passed a length of pale yellow segmented tube. I thought it could be part of a tapeworm.

"No, it was probably just a section of your bowel lining" my GP suggested when I got home.

"Just!"

As a strict Muslim town, Gorontalo offered no meaningful nightlife, at least not until Kennecott's expatriates had been there a few years and created a market for bars and eating places. We were pretty much the only "Westerners" in town and were invariably the centre of attention whenever we ventured out, sitting prominently in the back of a bendi[29].

---

29 A covered cart drawn by a small horse

I spent as little time as I could in town, preferring the bush camps, which were healthier (no careless humans unwittingly breeding malarial mosquitoes), quieter (no muezzins calling the faithful to prayer) and more visually appealing, sitting as they did deep within the Sulawesi rainforest. It was also safer, as earth tremors were quite frequent in town and I was never very confident about the structural integrity of the old house. During one scary tremor I saw the walls buckle and bend as the quake waves passed along them, and yet the house remained intact, so perhaps my fears were ill-founded.

Whenever Kennecott was testing its copper prospects in Sulawesi by diamond drilling there was a helicopter on duty to supply the rig and move it from one drill site to the next. For us geologists that made for easy transportation from Gorontalo to the field sites and from one prospect to another. But drilling programs came and went and although a chopper was sometimes available between them for geochemical stream sediment sampling and geophysical surveys and the like, there were periods when we had no helicopter. In those intervals, our transportation was of a rather more modest style, travelling by ex US Army four-wheel-drive trucks to the more accessible destinations. For others, we used boats to travel 60km along the coast to Tombuililato village then simply walked inland.

This was generally successful, but things did not always go according to plan.

WATER

"We need to do some follow up stream sediment sampling at Cabang Kiri," announced Bob, the Sulawesi program manager. Bob Clynch was a tall, rugged American, with blond hair and a crew cut that made him look like an action movie hero. He was also a competent manager and geologist who ran the complex and challenging program as efficiently as the "Indonesian factor" would allow. At one stage I recall we were spending $1 million a month, a large sum today but even more so in the mid 1970s.

"Why don't you go along with Stuart? It'll give you a good chance to see Cabang Kiri and get an idea of what its potential really is," Bob proposed.

"Sure, sounds good."

Up to that stage I had been concentrating on the much closer Tapadaa prospect, which we could access from Gorontalo by four-wheel drive truck. We had a large field camp at Tapadaa, where the geologists were supported by a team of local village labourers, 570 of them in all. Their task was to dig trenches for sampling and to construct, by hand, roads and tracks for drill rig access. I always marvelled at the sight of 500 or more labourers filing off to work in the early morning, each man carrying his ration of 1 kg of rice per day. The rice was cooked for them by kitchen staff using 44-gallon drums, starting daily at about 3.00 am; canned fish added protein to the starch.

"You can go back to Tapadaa when you return," Bob continued.

"OK, so how do we get to Cabang Kiri?"

"You'll have to go by boat to Tombuililato and walk in from there. Cabang Kiri's about 15km in from the coast. Stuart's arranging for the transportation."

Stuart was a fellow geologist about my own age, bearded, with a stocky build and unflappable disposition. He was already very experienced when it came to field work in the Indonesian jungle so I was happy to defer to him for arranging the trip along the coast to Tombuililato.

***

"Right, it's all set up" announced Stuart, "we leave tomorrow morning at 3.00 am."

"Why the hell are we leaving so early?" I asked, more than a little peeved at the thought of the short night ahead.

"You told me the speedboat would get us there comfortably in 3 hours."

"Ah," replied Stuart, "there is a problem with Tofi's boat. I've had to make other arrangements. We're umm ... ah ... going by canoe."

"What? It's 60km from Gorontalo to Tombuililato, we can't go that far in a bloody canoe." I was incredulous but Stuart remained calm.

"It's not just a canoe," he explained, "it's a big, sea-going canoe with a 50HP motor attached. It'll do the job."

I was dubious but there seemed to be no option. We had to get to our destination the next day to rendezvous with and relieve our colleagues who were due for a break from the bush camp at Cabang Kiri. Stuart was the old hand, after all, and I was the relative neophyte.

***

The canoe was indeed large, essentially a hollowed-out log about 8 m long, with high sides and a transom built on the stern to hold the impressive outboard motor that hung off it. In the darkness I could not see what equipment was stowed within the canoe but I assumed these people knew what they were doing and climbed aboard after Stuart. The canoe had a crew of four Indonesian men, including the skipper, who sat at the stern and deftly steered the canoe down the river a short distance to its mouth and then out into the Molucca Sea. Once there, he opened the throttle and the canoe began to slide through the calm water with ease. There was a pleasant off-shore breeze and the sky was clear, with a full complement of stars, which were not quite as sharp and sparkling as they would be in Australia due to the humid air between us and them.

"This is impressive," I called to Stuart, making myself as comfortable as possible on the hard wooden seats, "I think I'm going to enjoy this trip more than I expected."

"Yeah, I haven't used these blokes before but they seem OK. We should be there in about four hours."

Famous last words!

***

Hardly had Stuart spoken those words, it seemed, when the engine stopped. At first I was amused – the Indonesian factor at work once again – but after several minutes of the skipper trying to restart the motor, amusement turned to concern. His efforts continued but were equally futile. Concern turned to anger. The off-shore breeze continued to blow and to me it seemed stronger than when we started.

"You'd better get them to start paddling," I implored Stuart, "I don't like the way those lights on the shore are getting fainter."

Stuart issued the appropriate instructions. A babble of rapid-fire Bahasa[30] came back from the boat crew, too quick for me to follow.

"Ah …. It seems that we don't have any paddles!" Stuart informed me, unnervingly calm.

---

[30] Bahasa Indonesia is the language of Indonesia.

"Shit! Now what do we do?"

Not only were there no paddles, there was no water and no food on board and the wind kept blowing. Anger turned to fear.

"The next landfall downwind is hundreds of kilometres away. We're in deep do-do!"

The lights on the shore began to fade one by one as we drifted in the darkness and more futile attempts to restart the motor continued. I could not really see Stuart's face in the darkness, but I was sure that it was as grim as mine.

Then, as we drifted in total darkness, we collided with something. There was a burst of staccato Bahasa.

"What's going on?"

Stuart quickly interrogated the boat crew.

"Believe it or not, we just bumped into some fisherman in a canoe."

For a few seconds I was stunned.

"Here we are drifting helplessly across the Molucca Sea in total darkness and you're telling me we just happened to bump into another canoe?"

"Yep."

"Bloody hell!"

As I pondered the unlikeliness of it all, my mind suddenly clicked into gear.

"Stuart, that canoe may be our last ever contact with civilisation. We've got to get them to help us."

Stuart needed no convincing. A few quick words to our boat skipper were followed by hollering across the water. It turned out that there were about half a dozen small one- or two-man canoes out there fishing in the dark. Several of them now approached us as our skipper explained our predicament. Negotiations over rescue fees began and continued for some minutes.

"What are they arguing about?"

"They're asking for outrageous fees to help us."

"Tell 'em we'll pay whatever it costs for them to tow us back to shore," I declared, my mind focussed very much on my own survival.

"No, we have to bargain the price. It's the Indonesian way," he replied, his experience ultimately proving invaluable in the circumstances.

A settlement was reached and lines were attached from three small fishing canoes to our much larger vessel. They began paddling. I did not like to think too much about the physical effort required of the fishermen in towing this big, laden canoe all the way back to Gorontalo. Slowly, the lights on shore began to grow stronger. Fear and anger subsided to mere annoyance.

It took a couple of hours, but finally we passed through the river mouth and continued upstream, just after dawn as the local people were coming down to the river for their ablutions.

What a sight we must have presented to them! Here we were, two white bosses sitting up boldly in the disabled canoe while being rescued by a few humble fishermen. Annoyance turned to embarrassment as we reached our starting point and stepped ashore. Stuart went off to seek help from our base while I waited by the wharf, feeling very conspicuous.

***

Two hours later, Stuart returned.

"I've got us another boat," he announced, "Tofi's cousin. I'm assured its very reliable."

"Better bloody be!"

Within the hour the boat appeared. Meanwhile, I had been to a shop nearby and bought plenty of bottled water and something to eat, just in case.

Three hours after that we arrived in Tombuililato, following a pleasant but uneventful trip along the coast.

"Was I dreaming or did we nearly end up lost at sea this morning?" I asked Stuart.

"Nah mate, we simply experienced yet another example of Indonesia's rich and diverse culture!"

22

SUNGAI ULAR BESAR

AIR

The helicopter begins its slow descent towards our destination. It is forty minutes since we left the security of civilisation at Gorontalo, rising through still air in hot sunshine and heading east towards the cloud-misted mountains that form the backbone of Sulawesi Utara[31]. Suspended in a tiny machine that is dwarfed by the landscape around us, I feel very exposed. As we leave the bright sunshine and continue cruising just below the cloud base I am glad that Visual Flight Rules are mandatory up here, where clouds can have solid centres and there is no flat ground for an emergency landing. Helicopter flying is a tussle between exhilaration and fear, especially here. It seems unnatural, it is unnatural, but for us it is also the most practical way to penetrate this wilderness. Lifting off my headphones I listen once again to the loud, reassuring whine of the Jetranger's turbine engine. The dangers are many, and they are not just mechanical, as I look to the southeast and see a build up of huge cumulonimbus thunderheads, linked to earth by a grey veil of heavy rainfall. Heading this way, no doubt.

---
[31] i.e North Sulawesi

Beneath and beside me the land is draped in a smooth carpet of green, unbroken except for the occasional grey cliff or yellow landslide scar. Descending further, the carpet begins to resolve into individual trees, giving it texture, more a shagpile than a Berber. Here and there Kapok trees protrude above the canopy, distinguished by their straight vertical trunks and horizontal branches, with sparse leaves and dangling seed pods. Now we have levelled out, flying parallel to the heavily wooded slopes on our left. How can such large trees stay upright on that steep ground? Now I can look into the crevices that split the verdure and see, deep within them, ribbons of white that mark the runoff from yesterday's rain. Our destination appears, ahead and below, a helipad that is nothing more than a flat postage stamp dug by hand into a spur on the steep bank of the creek. From here it looks miniscule, treacherous, a tiny target almost lost amongst the grandeur of the mountains. Can we really get down there? Will this machine even fit on it? I guess I'll know soon enough.

The pilot executes a tight banking turn over the helipad, slowing our descent almost to a hover as he faces the helicopter out towards the open sky. The constant whine of the turbine is replaced now by a fast and rhythmic whup whup whup, as the spinning rotor tilts and cuts through the exhaust from the engine. Gently, under total control, the helicopter settles onto the helipad. Yes, we do fit, just, and we have arrived safely at Cabang Kiri. Marvelling yet again at the skill of these ex-Vietnam American pilots, I jump out of the chopper, running with lowered head to the edge of the helipad, where John shouts a greeting into my ear. We stand nearby as, without shutting down, the machine is quickly unloaded by two of the field assistants. John is calm while I am anxious as I keep an eye on the spinning tail rotor. I know I shouldn't worry; these guys are experienced fieldies, but the hazard is real and it makes me nervous just to watch. The task is soon completed and John gives the pilot a thumbs-up. Without further ceremony, the helicopter lifts off, rising vertically at first, then dipping its nose and accelerating forward, fleeing to the open sky where it turns right and quickly disappears from our sight and hearing. My ears ring in the silence that follows and I feel an unsettling sense of abandonment.

***

My role as a research geologist for Kennecott had once again taken me to Gorontalo in Sulawesi Utara. By this stage in the programme, drilling was underway at a prospect dubbed *Cabang Kiri* (Left Branch), situated in the remote, uninhabited and densely forested mountains 60km to the east of Gorontalo and from 10 to 15km inland from the coastal village of Tombuililato. While the drilling was taking place we had a helicopter to supply the rig, its crew and the geological team supervising the drilling, as well as to move the drill rig, piece by piece, from drill pad to drill pad. That made it much easier for us itinerant geologists to access the prospect, by simply catching a ride on the chopper when it was making one if its regular supply trips from Gorontalo to the field site.

"That certainly beats walking in," I remarked to John, soon after getting off the helicopter.

"The first time I came here it was by foot from Tombuililato."

"Yeah, I know," John agreed, "I've done it that way, too."

A temporary field camp had been built at Cabang Kiri to accommodate our crew. Typically, these camps consisted of a kitchen, a mess to eat in, doubling as an office at night, sleeping quarters and an ablution area, all constructed using a framework of timber that was covered with rolls of plastic sheeting. The timber usually comprised saplings cut from the forest and the rolls of plastic were flown in by chopper. They were quick and cheap to build and were readily abandoned when it came time to move on. (The plastic soon disintegrated under the dual attack of sun and rain.) Some of the longer-term camps became quite substantial, with several separate structures. Pathways were built between them, edged with white-painted stones, and generators were brought in to provide light and power, using petrol also flown in by helicopter. Other short term camps were much more basic, little more than a plastic shelter shed in the middle of the wilderness. The camp I came to that day at Cabang Kiri was pretty much in the latter category.

The stream sediment programme that had been underway during my first visit to this prospect, made laboriously by hiking in through the jungle, had pointed to a very attractive copper target for drilling. John, my senior geologist colleague, had been running the drilling programme testing that target from this camp for several weeks, accompanied by another geologist, Stuart, my erstwhile guide. They were supported by a dozen or so Indonesian field assistants. My visit was intended to assist John's geological interpretation.

"I want to show you some outcrops I've found further up this creek," he said, as we stood beside the helipad and peered down into the dim depths of the gorge below us.

"I think they're a good exposure of the porphyry copper system we're looking for here. I'm hoping your petrological expertise will help me decide if I'm on the right track," he continued.

"That's what I'm here for. But, first things first, I need something to eat. Breakfast was damned early today so I'm feeling a bit peckish."

While the field assistants moved the new provisions I had brought with me on the helicopter from the helipad to the nearby camp, John and I grabbed a snack and a drink. Next, I watched as he instructed the fieldies to go back up the slope to the ridge above us, where Stuart and the rest of the fieldies were working near the rig. The field assistants were hand-excavating contour trenches into the saprolite (soft, deeply weathered but *in situ* rock) so that John and Stuart could map and sample the exposed geology.

"Jeez John, your Bahasa has improved out of sight," I said admiringly, as the fieldies set off.

"Try living with them in the bush for weeks at a time – it's a matter of survival."

Once the fieldies had departed, John and I scrambled down the steep bank into the creek bed below and began to climb upwards along it, towards the intriguing outcrops.

"It's bloody dark down here John, how the hell am I going to see anything?"

"You'll get used to it," was his dismissive reply.

***

WATER

The creek bed is the only more or less unvegetated place in the forest. As such, it gives us a challenging but workable pathway through the jungle. We are in a dark, damp and rocky gorge, where the air is still and tainted with the fetid, heady smell of rotting vegetation. The metal cleats on our boots leave scrape marks in the moss on the boulders as we climb up the steep drainage. Leeches, hungry for our blood, arch into omegas on low slung leaves as we pass. Creepers and vines hang down, as though especially trained to impede our progress. Vicious tendrils of rotan, with their two-way thorns, catch our clothing and tattoo our arms. Again and again we scale waterfalls and cascades or scramble through the forest around them. We offer advice to each other as we climb –

"There's a foothold here."

"A handhold there."

"Watch out, that rock is loose."

"Mind that bloody rotan."

"There's a leech on your back."

Our voices echo off the rocky walls, sounding ethereal, seemingly an octave above their normal pitch as they float above the roaring torrent.

We wade through pools and climb over fallen trees or under the large stag horns that cling to the rock in these dark recesses. I feel that we are being watched, resented, by the diverse life that lies just beyond my vision. What manner of evil lurks in the crown of yonder tree fern? The mid-pitched drone of cicadas is white noise emanating from the green sponge that surrounds us. The more humid it becomes the louder is their rhythmic beat. A large hornbill suddenly takes flight in the canopy above, causing my heart to skip a beat. The pneumatic throbbing of the bird's wings makes it sound much larger than it is.

Our boots are wet; our clothes are wet; everything is wet. And then it begins to rain. What's the time? Quarter to one. You can just about set your watch by it in these parts. The light fades even more as the forest surrenders to the warm downpour. The roar of water in the creek competes with the intimidating thunder that is heavy rain in a tropical rainforest. All other sound is lost in the fluid commotion. The creek level rises rapidly. Soil in the banks turns to mud. Rocks become even more slippery. Ouch! I bang my shin again as I lose my footing for the umpteenth time. It is hard to tell what is rock, what is soil and what is simply suspended moss or rotting leaf litter. The leeches love it. What do they feed on when I am not here? Progress is slow, but we persevere.

***

We had been climbing up the watercourse for about an hour and were getting close to our objective. As we stopped for a spell I asked John why this creek was called 'Sungai Ular Besar[32]'.

"I was climbing up here one day when I came across a giant

---

[32] 'Sungai Ular Besar' means 'Big Snake Creek'.

Sulawesi python that had captured a wild pig and was squeezing the life out of it as I watched."

He recounted how with every porcine squeal the python tightened its grip. It was clearly a one-sided contest and there would be no escape for the pig.

"I didn't wish to interfere with the course of nature, so I struggled up out of the creek bed and into the adjacent forest to give the python a wide berth," he continued.

"Wow! Did you get a photo?"

"No, it was pissing with rain then just like it is now. It was too dark and wet to think of photography."

"I can quite understand," I rationalised, wiping the drips from my eyes as I surveyed the surrounding gloom.

"It's not the first time I've seen a python in the forest," responded John, "but it's certainly the most dramatic encounter I've had. Hence the name."

We pushed on up *Sungai Ular Besar* until finally we reached the place John had wanted me to see. In what seemed like a masterpiece of organisation on his part, the rain eased just before we arrived at our destination. For a couple of hours we studied the outcrops, discussing the significance of the rock types, alteration and traces of mineralisation we could see there. At intervals, we chipped off samples of the rock with our geological hammers and placed them into numbered sample bags for me to take back to Gorontalo and send to the lab for analysis and petrological study.

The humidity was intense but ...

"At least it's not raining," I declared.

"Just wait," said John.

"Yeah, I did see some pretty awesome thunderheads as we flew in."

Sure enough, soon after we began our return to camp the rain resumed, seemingly heavier than ever.

"How can so much bloody water fall from the sky at once?" was my sodden query.

We continued downwards. The descent back down the creek was arduous and, in those conditions, quite dangerous. Climbing down a waterfall is more difficult than climbing up. You cannot see where you are putting your feet, hand holds crumble, the rocks are slippery and everything is wet, wet, wet! As we neared the camp we found ourselves stalled part way down a cascade with a three metre drop still to be negotiated. There was no obvious way to climb down.

"How the hell did we get up here?" I asked.

"Buggered if I know," replied John, "I guess we must have climbed around it through the bush."

Everything looked different on the way down.

"Ah to hell with it," I said after a little while, "I'm not going to climb back up this thing. Here goes."

I couldn't possibly get any wetter so, taking a deep breath, I jumped off the little cliff, straight into the plunge pool at the bottom. John looked startled but then quickly followed suit. Soon after that two very wet and bedraggled geologists made it back to camp.

The downpour continued until dusk, making our mealtime miserable, the fieldies grumpy and John, Stuart and I tense. Finding a leech between my toes when I removed my boots did not help my mood, especially as it had ruptured and it was my blood that coloured my dripping socks.

Finally, the rain stopped and we could even see stars in the sky. But then the wind came up. It is rarely windy in this region, being so close to the equator, and the forest is not well adjusted to it. That night it sounded like a gale, though we were sheltered enough down in the camp and our awareness of the wind came mostly from the noise it made in the tree tops.

"This is the bit I don't like," John said, as we prepared to settle down for the night, "wind after rain is when trees fall down."

***

It must be after midnight. I've been asleep but something has woken me up. John is restless too, while Stuart snores contentedly. Not a scintilla of light penetrates into our hut. I have a torch here somewhere; should I turn it on? I lie awake listening to the wind in the canopy, imagining the trees straining to hold their footing. It is the only sound I hear; it seems ominous, menacing, as though arboreal ghosts are howling at our intrusion. All else is unnervingly quiet; even the usual chatter of nocturnal insects is absent. Are they as intimidated by the gale as I am?

Suddenly I hear a loud crack very close by. John and I both sit bolt upright – we know immediately what it is. Stuart stirs but does not wake. The falling tree begins to crash through the adjacent forest, no doubt wreaking havoc on smaller trees around it. Instant decision time – is it coming for us? Do we run from the hut in the total darkness and perhaps run right into its path? Or fall into the adjacent gorge? We have seconds to decide. We sit tight. Thud! The ground shakes – the tree is on the ground but it is not on us. We have survived!

***

"Jesus, that gave me a fright," said John, as we lay awake in the darkness afterwards, hearts still pounding.

"Bloody hell John, how often does that happen?"

"First time I've had one come that close," he replied, adding "and I hope it's the last time too."

"I'll drink to that."

Stuart had woken at the thud and asked what was happening.

"A bloody great tree has fallen close to us but we seem to have escaped damage."

"OK, cool. Goodnight."

I envied Stuart his composure as he quickly resumed his slumber.

My sleep was fitful at best for the rest of that night as my mind relived our frightening experience. I realised that if the tree had hit us no one would know about it until the return of the helicopter that was due in a few days to deliver more supplies and pluck me out of this remote and hostile environment.

The next day the wind had eased and our surroundings looked quite benign in the early morning sun. The terror of the night seemed almost imaginary.

"Let's see if we can find that bloody tree," I suggested, "I'd like to see just how big it is and be sure I wasn't dreaming."

"It was no fucking dream," John replied, in language that was unusually colourful for him. He was clearly still as rattled as I was.

For the next hour, all three of us scoured the forest near the camp, looking for a recently fallen tree. Surely it would be easy to recognise. The thud when it hit the ground was huge – it was certainly no sapling! But try as we might, we could not identify the culprit. The hungry rainforest had swallowed up one of its own and we would never know how close to death we had come.

***

It is nearly a year later. I am again with John in the air over Sulawesi Utara, being flown to a new prospect at Kayubulan Ridge. He wants to call into Cabang Kiri on the way to the new site to pick up some rolls of plastic he left behind when departing from the camp last year. John is in the front seat, next to the pilot – another Vietnam veteran. Once again the helicopter descends towards the old camp with its perilous helipad. As it does so, John's voice comes over the intercom:

"Jesus Christ!"

From the back seat I lean forward to see what has prompted this uncharacteristic outburst.

I see it and my heart skips a beat.

"Oh my God!"

No other words are adequate, as we both stare at the giant tree that has fallen right across the abandoned camp, completely demolishing it.

23

KAYUBULAN RIDGE

EARTH

Our quest for copper in Sulawesi continued, with the focus having shifted to a remote, uninhabited patch of jungle some 25km inland from the coastal village of Tombuililato. The village was located on the south coast of Sulawesi's North Arm (*Sulawesi Utara*), about 60km east of the city of Gorontalo, where we had our base. The plan was simple. Regional stream sediment geochemical sampling conducted earlier had identified this area as anomalous in copper and gold. Our job was to identify the source of the metals in stream sediments and assess the area's potential to host a major copper-gold deposit. We were just two geologists, John Dow and me, assisted by about one hundred Indonesian labourers from coastal villages, who had walked in from Tombuililato ahead of us to set up camp beside the creek and clear a helipad.

For John and me, getting there was much easier. Early one morning we left Gorontalo by helicopter, a snappy little Hughes 500. After an hour or so the chopper was circling above the camp site and adjacent helipad, which were alternately hidden and revealed as a bank of morning fog drifted in and out of the narrow valley.

"Do you really think we can put down there?" I asked the pilot, expecting him to say no.

"We'll give it a go. Should be okay."

Skilfully, he turned the aircraft in ever tightening circles and dropped slowly down towards the landing site.

The fog continued to obscure the helipad as we descended between the trees on either side of the creek valley. I was nervous; John was a bit unsettled too. But the Hughes 500 had one advantage over the more commonly used Jetranger[33] – it had five shorter blades (spinning diameter of 8m) instead of two longer ones (spinning diameter of 10m). To my amazement, the pilot banked and spiralled the helicopter almost in one spot, with the trees just out of reach of the spinning blades. As he did so, the downwash from the rotors blew the fog away just long enough for him to drop the aircraft down onto the helipad.

"Jeez, I didn't think we'd get in today," I remarked to John.

"Never underestimate what these jet-jockeys can do," John replied, speaking from experience.

Within minutes the helicopter had been unloaded and John and I were gazing up at Kayubulan Ridge, rising steeply above us. Further supply flights would take place at least weekly while we were there, delivering food, including vast quantities of rice and canned fish for the men from Tombuililato, and petrol for the small generator that would give us light and power our little refrigerator. It was very basic, comfortable enough in the circumstances, but like it or not, this remote, primitive field camp was to be home for John and me for at least the next month. The camp had been built before our arrival on a small patch of flattish land beside a deeply incised creek. The water in the creek had a pH of 2.2, essentially dilute sulphuric acid, or not so dilute really. This was due to the abundant pyrite (FeS_2) disseminated throughout the local rocks. As those rocks weathered

---

[33] Bell 206 - "Jetranger"

at surface, the pyrite was oxidised, releasing sulphate into the water, strongly acidifying it.

Our exploration target loomed high above us and each day began with a gut-busting, zig-zagging climb up the seemingly vertical slope to reach the main prospect site. Kayubulan Ridge comprised a large body of quartz diorite porphyry (an igneous rock, similar to granite) that had been intruded millions of years ago at a depth of 1 to 2km below the then surface. As it cooled, the rock was soaked by hydrothermal (hot water) fluids left as residue from the crystallising igneous source. The primary igneous minerals had been altered by the fluids to an exotic secondary mineral assemblage. More importantly, those same fluids had contained copper and gold, concentrated into the residue and then deposited as free gold and chalcopyrite (a copper sulphide mineral) in veins, fractures and disseminations through the host rock. Over time, erosion had exposed the mineralised rock at the present day land surface and made it our target. Atop the ridge sat a cap of hard, siliceous rock[34] that was resistant to erosion. Down the flanks the rock was softer, rich in clay and deeply weathered – it is not only organic matter that rots quickly in the humid tropical climate. Our job was to assess which part of the ridge carried the most copper and gold and thus offered the best chance of an economic deposit.

Over the next month, John and I mapped and sampled the entire prospect in sufficient detail to allow proper evaluation of its potential. The assessment was complicated by the fact that the gold was never visible because it was very fine grained and the copper, once oxidised, was very vulnerable to leaching in the relentless tropical rain. Very little copper remained in the exposed surface rocks. For this reason, we had to take a great many samples for assay, particularly for gold, as gold was not leached in this environment and its distribution was a

---

[34] Advanced argillic alteration for those who are interested in such things.

good guide to where, at depth, the best copper might be found. The dense forest cover, steep slopes and deep weathering meant that to obtain reliable, *in situ* samples we required the labourers to excavate, by hand, long contour trenches around the mountainside. With a couple of the villagers as assistants, John and I formed two teams, recording details of the rock exposed in the trenches and collecting samples for later chemical analysis.

"*Ambil sampil disini?*" (take a sample here) was my constant instruction, as I scratched a horizontal, two-metre-long groove in the back of the trench wall with my G-pick[35] while the assistants held a bag beneath to catch the loosened soft rock.

It was hard, back-breaking work in the sweltering tropical heat, and not without its dangers:

"I was sampling in a trench one time when it collapsed in on me," John explained, "burying me up to my chest."

"Sounds pretty scary. How did you get out?"

"Well, would you believe, the bastards started wailing and lamenting my loss instead of digging me out!"

It appeared that John had to use every Indonesian expletive he could think of to motivate them to dig him out before further collapse.

"It wasn't that long after we had actually lost a labourer who had ducked when the wall of a trench he was digging collapsed. He was buried under the debris, so I was very conscious of the threat."

So, dangerous it was but also strangely exciting, as we explored this previously unknown deposit of copper and gold. At least we became fit, very fit, in the process. And we were productive. Every visit from the helicopter saw us load a couple of hundred samples on to it to be flown back to Gorontalo and submitted to the laboratory, where they

---

[35] Geological hammer, with a blunt end and a pointy end.

were analysed for copper, gold and a host of trace elements.

Steadily, we accumulated the information that would be needed to assess the potential of the area. Both sides and the top of Kayubulan Ridge were sampled systematically, eventually covering several square kilometres. The days were long and repetitious but we knew the harder we worked the less time we would have to stay there. At the end of each day John and I plotted the results of our work onto maps that slowly began to fill with data and guided our plans for the next day's efforts. Day by day the gaps were fewer and it looked like we would complete the program as scheduled.

After work each day the Indonesian labourers bathed in the adjacent stream, despite its low pH, and then relaxed for a short while, smoking their kretek cigarettes[36] and talking animatedly amongst themselves. They had a separate camp nearby, constructed from forest poles and covered with plastic. Their evening meal, mostly rice and canned fish, was generally eaten before dark and they retired soon afterwards. For a while a murmur of Bahasa conversation would rise from their camp, soon fading as exhaustion overtook them. Then the only indication of their presence was the aroma of cloves drifting across as they smoked their last *kreteks* for the day.

Our camp was just as primitive as that or our workers, made with timber from the forest and covered with plastic that came in long rolls. For washing we had the luxury of a shower bag hung from a tree and filled on request with hot water by the camp cook. We also had a table to sit and eat or work at. It, like the benches we sat on, was made of poles cut from the forest and covered in split bamboo, functional but hardly comfortable. We became immune to the sound of the generator drumming away in the background as we sat under

---

[36] Particles of clove are mixed with tobacco in this distinctively Indonesian addiction.

the glare of the single bulb dangling from the roof and devoured the filling but rather repetitious meals served up by our cook. He had a multitude of cans to work with but a very limited stock of fresh food and did a good job in the circumstances. We were so hungry by the evening we hardly cared.

There was no communication with the outside world so we were dependent on each other for entertainment. After eating, our conversation ranged widely, as we talked of family, career ambition, our chances of success and a thousand other subjects. It was October and one night John, who was a keen follower of horse racing, talked for an hour about all the probable entrants for the Melbourne Cup and who might be the winner. It would be weeks before we knew if he was correct but I found it entertaining just the same. Some nights we were untroubled by insects, while at other times the bugs flooded in from the forest. One night the moths formed a swirling, brown, almost solid cylinder, a half metre high and rotating rapidly beneath the light bulb. When we finally retired and turned the generator off, the silence was initially absolute, but then, as our ears stopped ringing, the sounds of the forest at night began to emerge, lulling us to sleep, content with the knowledge that we were one day closer to going home.

***

The work was almost done; a few more days would finish it. Unfortunately, however, those few days extended beyond the life of the helicopter contract, which meant that we would have to walk out to the coast rather than fly comfortably back to Gorontalo. With the last supply flight due the next day, John and I discussed our options.

"It's very tempting to say we've done enough, John," I proffered, unsure myself whether I really wanted to cut and run, or rather fly.

"I know, but we're here to do a job, so we might as well finish it, even if we have to walk out." John was his usual thorough self.

"And at least we're fit so the walk shouldn't be a problem."

"We're fit enough for sure, but we will have to leave much of the camp gear here, won't we?"

John, ever the committed exploration geologist, was resolute.

"Better to leave replaceable stuff behind than to leave part of the prospect unmapped and unsampled."

The next day everything we figured we could do without and every sample we could muster was sent out on the last chopper flight. A request for a pick up by boat in Tombuililato a week later went out with the helicopter.

We had been feeling isolated over the previous month but now, with no helicopter to come to the rescue, the sense of isolation and abandonment was complete. It was not that we were scared or worried. We were quite confident in our ability to walk out. But now the remoteness invaded our consciousness more deeply and made us more aware of our surroundings. The mountain seemed steeper than ever, the jungle thicker, more obstructionist, and the daily labour more gruelling than before. The canned food served up each evening lost all flavour and our palates began to savour the luxuries of the fresh vegetables and juicy, rare steak that awaited us back in Gorontalo. But we had to finish the job first so the field work continued for several more days.

Finally, we were satisfied that we had covered the whole area properly and there would be no need to return unless the results pointed to a serious target for follow up[37]. On our last day in the

---

[37] The copper deposits we found were never developed and the area was later included in a new national park.

camp, preparations for the trek to the coast were made. John and I would carry nothing; our personal field assistants would carry our backpacks. The remainder of the one hundred or so villagers would each carry 15-20 kg of field gear or samples so that nothing of value was left behind (the fridge and the generator were left in the camp, perhaps to be retrieved in a future campaign).

At dawn the next morning the column filed out of Kayubulan camp, carriers hefting their loads while John and I, unburdened, led the way. It was like a scene from an old Hollywood movie. At any minute, you might have expected Tarzan to swing by, or at least to hear a safari-suited John Wayne mouthing banalities. But no, it was an Aussie geologist and his Kiwi colleague – no lights, no cameras, but plenty of action.

***

The route would take us south along Kayubulan Ridge, which was on the northern side of the main range. We would then follow a rudimentary track along that range west for some distance before taking a ridge on its southern side that would lead us down to the village of Tombuililato on the coast. The total distance was about 25km. Fit as we were, and unburdened by having to carry anything, we started to draw ahead of the main gang of fieldies, who struggled uphill with their loads. Before long we reached the main range and turned west along a knife edge ridge. Knife edge is no exaggeration; in places the track occupied the entire width of the ridge and I could touch trees metres above their base by just reaching out sideways. John and I continued briskly, drawing ahead even from our personal carriers, who were also our guides. I was upbeat:

"Not to worry. The track seems well marked enough. We might as well keep going so we can reach Tombuililato by dark."

We found the southern ridge where we expected it and turned towards the coast. That is when things started to go wrong.

Initially the track was still quite evident and seemed to follow pretty much the very top of the spur ridge. But after some time we came to an old helipad, where many trees had been cut down and now lay haphazardly around the clearing. Following the ridge, we walked into the clearing with confidence, which soon evaporated as we could not pick up the track on the other side of the helipad due to all the fallen timber.

"This looks like it," declared John finally, as we climbed over the debris and started along a relatively clear path through the forest.

"OK with me." My confidence returned.

Ten minutes later the apparent path petered out as the scrub closed in around us. My confidence fled.

"Shit! Now what?"

"We might as well stick to this ridge," suggested John, "at least it's headed in the right direction."

The jungle around us was as rich and diverse as any rainforest in the world and we were well used to the thick understory that grew beneath the canopy. There was always something to look at, some huge tree that defied gravity, an exotic plant or flower to admire, a hanging vine to hold on to. But there was one plant that we all detested: In Australia, we know it for the sturdy canes it produces that are used to make cane or "rattan" furniture. The Indonesians know it as "rotan".

Rotan is a particularly nasty plant. The thicker tendrils, the ones used to make furniture, trail from their leafy roots for many metres and are covered in a scaly sheath that bristles with hard, sharp thorns up to several centimetres long. Between these larger spines are smaller ones that point in the opposite direction, so there is no escape: the plant gets you coming and going. The large serrated leaves themselves terminate in trailing finer canes, armed with small curved talons that scratch and rip skin with ease. It is a bush to be avoided at all costs. But sometimes that is just not possible.

"Watch out John, there's a bloody great rotan just ahead."

As if John couldn't see it!

"We can try diverting off the ridge for a bit," he suggested, "maybe get round this rotan patch and back up on to the ridge top a bit further on."

"Worth a go."

For thirty minutes or so we scrambled through the scrub on the steep side of the ridge, continuing, we thought, in roughly the right direction. But it was not.

"I dunno, we must have got off onto a side spur somehow," John decided, as the ground we were on started to drop precipitously down towards the river, far below.

We knew it was much too soon for that to be happening if we were on the correct ridge.

"Yeah, we should be headed south but I don't think we are."

John agreed, and added:

"But I'm not going backwards so we might as well push on. We'll try to stay high as we can."

It soon became apparent, however, that we were in trouble. It was not that we were lost, not ultimately; it was just that we did not know where we were. We knew that the river, way below us on our left flowed into the sea at Tombuililato and that eventually we would need to descend into it. But we also knew that in its upper reaches the river was impossibly steep, with numerous treacherous waterfalls. We had to stay high.

"We're just gonna have to find our way back up to the main ridge again." John was resolute.

Reluctantly, we turned and went upslope. After some time and the extrusion of a great deal of sweat, we did come upon the main ridge again – right in the midst of a horrible thicket of rotan.

"I'm not leaving this ridge again," I announced, "rotan or no fucking rotan."

Our task might have been easier had we thought to carry a machete with us; our assistants had them but they were far behind us somewhere:

"Probably on the right bloody track!"

We were beginning to rue the fitness that had put us so far out in front.

The rotan thicket was too dense to walk or push through, but by lying on the ground we found we could wriggle along under the canes and dodge most of the vicious spikes. Progress was slow. Conversation was at a minimum. All effort went into survival.

After some time – ten minutes, an hour, who knows? – I suddenly heard a loud screeching just above me.

"Jesus Christ! What's that?"

"They're macaques," John called.

"Are they dangerous?"

"Not really."

"Oh yeah? What about that big bugger up there?"

I was looking up through the rotan to where the tribe of monkeys had gathered around a big male in the trees just above us. He looked as big as an African baboon. We had apparently surprised them by arriving on the scene prostrate on the ground. Clearly, any animal that crawls through rotan must be dangerous, so they started pelting us with fruit and bits of branches from the trees. It was not as though we could stand up and defend ourselves. The big one – the alpha male – swung down till he was just two or three metres above me. His mouth was a yawning cavern, framed by the largest canines I had ever been that close to. We continued our slow progress beneath the rotan; the monkeys kept pace with us above and continued their verbal abuse. Then a funny thing happened; funny in retrospect at least: The alpha male fell out of his tree and into the rotan right next to me. For an instant, we looked each other in the eye. Then his screams intensified – the thorns obviously hurt him as much as they did me. Quickly the large monkey scrambled up out of the thicket and led his tribe away, deciding to leave the spines and canes of rotan to those stupid wriggling intruders.

***

It was just on dusk when John and I walked into Tombuililato, savouring the delectable perfume of flowering coffee trees as we walked through the village gardens. The villagers greeted us warmly as we made our way to the rendezvous at the beach. Despite our forced detours and wild encounters, we were still ahead of the field assistants carrying our personal stuff and the other villagers carrying the residual samples and field gear. We sat exhausted on the beach, chatting to the *Bupati* (village chief) and drinking the refreshing juice of green coconuts that the village had offered us. Our base in Gorontalo had sent a half-cabin speedboat to collect us and the vessel and its driver were waiting patiently on the beach.

"Let's go," said John, finally, "It'll be a slow trip back in the dark and I for one am looking forward to a decent meal and a soft bed.

The boat can come back tomorrow for the samples and field stuff."

"I'm ready, but look, who's this?"

Out of the gloom a tall Indonesian appeared, carrying a backpack and looking just as buggered as we were. It was my personal assistant and I welcomed him with profuse praise for making it out in one day. I made sure that he, and the villagers standing around us, knew how pleased I was to have my gear to take with me and thanked him for his effort. He swelled with pride, seeming to grow even taller. To emphasise the point, I reached into my pack and brought out a neatly wrapped Villiger Export cigar, which I presented to him with due ceremony. As the speedboat began to pull away from the Tombuililato shoreline I studied his face: Never had I seen a man so exhausted but so proud all at once.

Or maybe he was simply struggling not to cough as he puffed away on my expensive cigar.

An Alouette II helicopter, with John Dow as passenger, sits on a helipad in Sulawesi, ca. 1976

4. THE PASSIONATE PROSPECTOR

24

NO STONE UNTURNED

AETHER

From 1980 onwards my working life took on a somewhat less adventurous but nonetheless satisfyingly diverse character, as I began an ascent up the management ladder and changed both my employer and the kind of employment several times. It began when I joined Pancontinental Mining Limited ("Pancon") as a Senior Geologist. My role was to lead an exploration team whose objective was to take the company away from its then primary focus on uranium (Pancon owned the undeveloped Jabiluka uranium deposit in the Northern Territory) and into gold and base metals.

We were successful, discovering and subsequently developing the Paddington Gold Deposit, near Kalgoorlie in Western Australia and later the Wodgina Tantalum Deposit in the Pilbara. Over more than ten years, from 1980 to 1991, I enjoyed considerable progress with Pancon, rising through the ranks from Senior Geologist focussed solely on exploration to Group General Manager with broad responsibility for Corporate Development, Exploration, Marketing and Human Resources. Along the way, in 1986, the company sent me, at its considerable expense, to undertake the three-month-long Advanced Management Program at the Harvard Business School.

My time at Harvard was the pivotal experience in my working life, marking the transition from "worker" to "manager". It began in late summer and finished in early winter, just before Christmas. Celibacy was a requirement (yet again) but partners were invited to join their students for a week at the end. Margaret joined me in early December, 1986, as the class moved from on-campus accommodation at the Business School to the Ritz Hotel in, by then, snowy Boston. The course was taught primarily using case studies written by Harvard professors and their students about real companies and real people in the business world. Initially I found it hard to relate the experiences of people in manufacturing or financial and service companies to my own situation in a primary industry (i.e. mining). Over time, however, and in retrospect, I began to see a few very basic and fundamental principles of business and economics emerge and they became ingrained in my thinking. I acknowledge that, whether I intended it or not, those principles guided my approach to work thereafter and were applied, to varying extent, in all my subsequent roles. One lesson, in particular, has stuck clearly in my memory: In a class dealing with interaction with the media, we had a visit from Morley Safer, of the US Sixty Minutes television program. His warning was direct and salutary:

"When Sixty Minutes comes visiting your company, it's never good news!"

My attendance at Harvard gave my life a balance I did not know was lacking until it happened. Four years at Berkeley had given me a superb start in my professional life and a strong belief in the efficacy of the top tier of the American education system. The University of California at Berkeley was and still is one of the great institutions of American education, with very high status, but it does not have the history or the aura of the Ivy League schools. Harvard, founded as a theological college in 1636, is the oldest institution of higher learning in the USA and is the nearest equivalent in America to Oxford and Cambridge in England. Indeed, the institution is located in the city of Cambridge, Massachusetts. One morning, after the first heavy snowfall of the season, I found myself wandering through Harvard Yard. Under a clear blue sky, the brilliance of a foot or so of fresh snow sparkled in the sunshine, covering the lawns and providing an artistic drape to the leafless trees. Across it all the bronze statue of a seated John Harvard stared imperiously from its snowy pedestal, anchoring this famous quadrangle to its place in history.

That experience was like that of my first visit to Cambridge, the original, very British Cambridge that is, many years before. In both places, the atmosphere of beauty, charm and history was palpable. The time at Harvard, brief as it was, gave me an indelible sense of achievement: I am a product of the best that American education can offer, in both its western and eastern incarnations. I trust that my subsequent life and career have reflected that privilege.

Not that such qualifications are any guarantee of success or security of employment. On the contrary, my tenure with Pancon ended abruptly in 1991 when the company experienced financial difficulty during Paul Keating's "recession we had to have". Costs had to be reduced and staff had to go. I was an expensive luxury (or so it was deemed) and my departure would cut the salary bill substantially. As a result, I was numbered amongst the victims of involuntary termination. But, as they say:

"You are not a real geologist until you have been retrenched at least once".

At the time, the experience was distressing and I remained without full time employment for two years after my retrenchment. But, as so often happens, it turned out for the best in due course. Temporary assignments and contract work kept the wolf from the door while I waited for a new opportunity to arise.

My period of underemployment finally ceased in early 1993, when I was appointed to the position of Director General of the New South Wales Department of Mineral Resources ("DMR"). This opportunity came quite out of the blue, following an approach by a 'headhunter' (i.e. an executive search firm). I was as surprised as anyone when I was successful. Not a moment's thought had been given on my part to the possibility of switching roles from industry to government but after two years of underemployment I was more than happy to give it a go. With my children still in full time education I was very pleased to take on a new challenge while receiving a comparatively modest but regular pay cheque.

My final work experience, before retirement from full time employment in 2012, was with a little company that I established from scratch – Malachite Resources Limited. Aided by some colleagues in the industry and friends in a major law firm, as well as several sympathetic seed investors, I set up Malachite in early 1997 as an unlisted public company, with a view to listing it on the Australian Securities Exchange ("ASX") within a couple of years. Soon after establishing Malachite the resources market had another of its cyclical downturns and my company struggled to survive as an unlisted entity. But survive we did and Malachite was finally listed on the ASX in 2002. Over the next ten years the company's fortunes waxed and waned, making a number of significant new mineral discoveries but never one that become economic and thus able to take us to producer status. Today the company continues to operate under new management and with new major shareholders and is operating a small gold mine near Cloncurry in northwest Queensland. The acquisition of this mine, known as "Lorena", was my last major action before leaving Malachite.

While all of that was going on I also served as a non-executive director (and for a while, Chairman) of Straits Resources Limited. Straits was formed to develop the Girilambone Copper Mine, near Nyngan in western NSW. I had officially opened the Girilambone mine for Straits early in my tenure as Head of the NSW Department of Mineral Resources (the only occasion on which I had this honour, which usually fell to a politician, either the Premier or the Minister for Mineral Resources). At "Giri", as it was affectionately known, the company mined the oxidised top of a copper sulphide ore body, where the primary copper mineral (chalcopyrite, $CuFeS_2$) had been converted by oxidation (weathering) over the eons to a mixture of chalcocite (Cu_2S), malachite ($Cu_2CO_3(OH)_2$) and other secondary minerals. The secondary minerals were soluble in sulphuric acid so the production method involved mining, crushing, stacking into heaps and then leaching those heaps with dilute acid. The copper in solution was collected into ponds from where it was pumped into a solvent extraction plant, where the copper was extracted from the acidic solution and transferred to a concentrated liquor that then went into an electro-winning tank house. There the copper was precipitated as cathodes by electrolysis, producing sheets of pure copper – 99.999% Cu. These were sold directly to fabricators, such as pipe and wire manufacturers. It was a very efficient process that was known for convenience as SX-EW copper production.

Straits later went on to operate other oxide copper (i.e. soluble copper) operations in Australia, including the Nifty Mine on the remote Great Sandy Desert of WA and then turned its attention to Indonesia. There (in Kalimantan) it very successfully developed a coal mine, bought a second coal mine and acquired a rather unsuccessful gold mine. In all I served as a director for 14 years, including three years while the company was based in Sydney (and I was Chairman) and eleven years after it moved to Perth. For me, that meant eleven years of almost monthly three-day trips to Perth for Board meetings and immersion in the company's activities. It was a long way to go but it did earn me a lot of frequent flyer points. Straits became a big part of my life over those years and there was much cross-pollination between Straits Resources and Malachite Resources.

These are but some of the many stories that could be told from the latter half of my life as a geologist, but I will spare you a detailed account of all those experiences. Instead, as a diligent geologist, I will just take samples from the "ore body" that is my later life, enabling you to form a view about the nature and quality of the entire entity. You will see, and I hope appreciate, how and why geology continues to motivate me and enrich my life.

25

SCIENCE vs SERENDIPITY

EARTH

As an exploration geologist, the question I have been asked by lay people more than any other is:

"How do you know where to look?"

That question pre-supposes that ore deposits occur randomly in nature; that mineral exploration is somehow a lucky dip or a lottery.

There can be a lot of luck involved in a successful mineral discovery but, fortunately for the likes of me, finding an ore deposit is more about science than it is about serendipity. Because mineral exploration has been such a large part of my working life, and pretty much made me what I am, I would like to give you a taste of what that science is and how it is applied to find a gold mine, or, indeed, any other rock of value to us.

The first thing to understand is that ore deposits are rocks too. They are not freaks of nature but the product of geological processes that are as much a part of the evolution of the earth as are volcanic eruptions, the deposition of limestone reefs, the building of mountains, or the intrusion of granite batholiths. The earth is quite indifferent as to the values we place on certain rocks or minerals. To find those valuable rocks and minerals, that we call ores, we need to understand the history of the earth, particularly the structure and processes at work in the crust and their evolution over time.

Exploration geoscience is the application of this knowledge in the quest to provide mankind with the energy and materials it needs to enjoy the quality of life that we have come to expect. It is well known how important mineral exports are in keeping Australia solvent. It can perhaps be easily understood that we need petroleum to make petrol and diesel, or iron ore and coal to make steel, or copper to make wires and pipes. But most people are surprised to learn just how much we depend on mineral products to sustain our living standards.

If you are a golfer, you need the mineral barite for your golf balls (it's the gooey paste in the middle). If you wear lipstick, you need iron ore to make it red. If you look in a mirror, flick an electric switch, or are treated for severe burns, you are using silver, which has more end uses than all other metals combined. When you wash your hair with a shampoo and conditioner you need silicones, made from silicon metal, which is produced from the mineral quartz. It is the silicones that not only keep your hair *'soft and manageable'* but seal up the leaks in the shower as well! A smart phone contains 18 different metals. To construct a modern power generation windmill you need nearly half the periodic table – 45 different metals, every one of which comes from a mine somewhere.

We are all major consumers of minerals and modern society simply could not exist without a continuous supply of these materials. Equally critical is the continual discovery of new sources of them to replace those that have been depleted by production. Although ore deposits come in all shapes and sizes, they are all finite and meeting society's need for useful rocks is the job of exploration geoscientists. In doing so they effectively perform the R & D[38] of mining.

---

[38] i.e. Research and Development

Exploration geoscience is a very broad church. Perhaps more than any other scientific discipline, it encompasses the application of knowledge from other sciences to aid the understanding and interpretation of our earth and its makeup. Hence, although geology has many sub-disciplines of its own, earth science also encompasses physics and chemistry in particular – what we call geophysics and geochemistry. It is in the measurement of physical and chemical parameters of the earth's crust that modern technology has been of immeasurable benefit to mineral exploration. The digital revolution has had as much impact on mineral exploration and mining generally as it has on our broader society and lifestyle.

Such modern aids are very welcome. The discovery of new deposits is becoming ever more difficult as we look deeper and deeper into the earth's crust. For, while we work on a 2-dimensional surface, we are looking for a target in 3-dimensional space. The top of that target, as in the case of the Olympic Dam copper deposit in South Australia, could be hundreds of metres below that 2-D surface.

Exploration is an expensive, high-risk process, especially once it comes to drill testing a prospect. Deep pockets are required. And that is before accounting for the cost of the 99 prospects that might have been evaluated and drilled unsuccessfully prior to the ultimate economic discovery. With such an expensive undertaking, good science, and a thick neck, are essential attributes of an exploration geologist! And being inherently lucky helps too!

In Australia, the days of stumbling over an outcrop of rock that just happens to be the top of a major ore body are pretty much gone. Probably never again will a lonely boundary rider, like Charles Rasp, stop for a rest on a hill and stub his toe on a heavy rock that leads to the discovery of a major ore body. In his case it was the Broken Hill lead-zinc-silver deposit, the greatest such ore deposit in the world.

Over its lifetime, which continues today, Broken Hill has produced, at today's prices, over two hundred billion dollars-worth of silver, lead and zinc. All of that has come from a remarkably small piece of land, just 500m wide and 7.5km end to end. As geologist and fellow elemental traveller, Ross Fardon[39] says, "mining is point-source wealth".

It is science, not serendipity, that is most likely to put us in the right place to find a gold mine, it is science that tells us where to drill and it is science that tells us whether what we find is actually what we were looking for in the first place. On the other hand, it is often serendipity that determines whether what we do find is "a right little gold mine" or just a "technical success" – yet another interesting but uneconomic mineral occurrence, adding to the vast inventory of such occurrences. In mineral exploration, like so most other walks of life, it is still better to be lucky than unlucky.

Right from the start we face long odds: In order of magnitude terms, for every 1000 mineral occurrences we might look at, 100 are worth more detailed work, of which 10 might yield a mineral resource and, if you are lucky, 1 will emerge as an economic mine.

The challenge for the successful explorer is to beat those odds, something I have done a few times, as I shall explain.

---

[39] Ross Fardon: *This Could be Your Future*. Xlibris, 2013, 389pp.

About 15kg of gold (worth nearly $1 million at today's prices), produced at Pancontinental's Paddington Gold Mine, ca. 1988

26

DISCOVERY

EARTH

In the late 1970's I was working as a consultant and contractor to mining companies across Australia, but with emphasis on NSW. My speciality was the application of petrology[40] to mineral exploration. A company called Geopeko (part of Peko-Wallsend Ltd) was one of my major clients and the source of my first big success. For some time, I had been providing petrological services to Geopeko, identifying and describing rock samples sent to me by their field geologists. One day a new batch was sent to me from work underway at Goonumbla, north west of Parkes in NSW. The rocks, I was told, came from shallow drilling along country laneways in an area with little or no rock outcrop.

"There's a bit of copper in these rocks," I was informed.

"We seem to have discovered some low grade volcanogenic[41] mineralisation."

Armed with an extensive knowledge of porphyry-type copper deposits from my days with Kennecott, I examined the rocks in thin section under my petrographic microscope and decided that the type of copper mineralisation they had discovered was not low grade

---

[40] Using my own Carl Zeiss petrographic microscope – a superb instrument and a joy to use.
[41] Associated with volcanic rocks and formed by volcanic processes.

volcanogenic but high grade porphyry style[42]. At first, they did not believe me but they did at least invite me to visit the site and see the local geology for myself. Soon afterwards, I spent a long Friday in the field with the local Geopeko geologists, pointing out to them characteristics of such rocks as did crop out that fitted with my porphyry diagnosis. At the end of the day, I was taken to dinner at a motel in Parkes, before being delivered to the Parkes airport for the flight home. Conversation was very animated during the meal and the food and drink were consumed without too much attention to detail, except for the garlic prawns, that were rather too garlicy I thought. Around 9.00pm I boarded the Fokker for the flight back to Sydney. There were only a handful of passengers, including me and three Geopeko personnel returning to Sydney for a break. The four of us sat abreast in Row 3. When the flight attendant came forward to give her safety spiel the first thing she said was:

"Oh my God! Garlic!"

When she had finished, the two nuns in the row behind us stood up and announced:

"Don't mind us boys, but we think we'll sit down the back!"

It was then that we realised the bed of "onions" under the prawns was not onions but garlic.

When I walked in through the front door at home near midnight, my wife called out,

"Is that you, Garry?" and then,

"Oh my God! Garlic!"

What Geopeko had found was but the top of a very significant new

---

[42] Formed within the upper crust when copper-bearing intrusive bodies of magma cooled and crystallised, often 1-2 km below volcanoes at the then surface, and later exposed by uplift and erosion.

style of ore body, previously hardly known in Australia – a porphyry copper deposit associated with volcanic and shallow intrusive rocks of Ordovician age[43]. The implication was immediately realised – vastly larger size potential and much greater metal value than they had previously thought. That deposit, or, as it turned out, that group of deposits, is still in production today and known as the Northparkes Copper Mine.

As the size and high value of Northparkes became apparent, Geopeko management decided it would be wise to look for any repetitions that might occur elsewhere in the state, before a competitor beat them to it. They knew that the Goonumbla deposits were associated with a very distinctive pattern in the aeromagnetic data, crude though that was at the time, reflecting the unusual volcanic rocks that hosted them. Looking at maps produced by the NSW Geological Survey they identified another area that displayed a similarly unusual pattern in the published aeromagnetic data. Geopeko applied for and was granted an exploration licence over that area, located north of West Wyalong. I was appointed to go and see what was there.

Over several months, I systematically mapped the geology of the area, except for the bit that was covered by Lake Cowal, an ephemeral lake that was filled with water at the time (like Lake George near Canberra, Lake Cowal is often dry). There were few if any known copper or gold deposits or old mines or prospects in the licence area so expectations of a successful outcome were not high. But I persevered, plotting the geology onto air photos and compiling it onto a master map. From time to time I took samples for assay, but not with much enthusiasm, as the rocks looked pretty "hungry". Then on one fateful day, all that changed.

---

[43] Around 450-470 million years old.

It was getting late in the day, a Thursday, and I was weary, having walked over a great deal of barren countryside during the previous hours, seeing lots of emus and kangaroos but no interesting rocks. For most of that and many earlier days I had been frustrated by lack of outcrop in the largely flat, soil-covered grazing country around Lake Cowal. By studying my air photos, I could see that there might be a bit of rock outcrop on the shores of the lake, several kilometres to the north and through many farm gates. It was tempting to skip it and return to base, keenly anticipating my return home due late the next day. But then the insatiable curiosity of the passionate prospector took over. I decided to go and have a look, just in case.

It took nearly an hour to drive the distance, such were the challenges of opening and closing farm gates and navigating along criss-crossing access tracks (all the work was conducted with the approval of the local land owners, to whom Geopeko paid generous compensation). Eventually I found myself at the shore of the lake.

To my great delight, there was indeed rock outcrop there. Not just outcrop, but exposures of rocks that looked surprisingly similar to the rocks that surrounded the Goonumbla deposits, further north. Encouraged, I started recording details on the air photo and in my notebook. Then I noticed that the nearby farm house, set back from the lake a hundred metres or so, was built on a slight rise. My attention transferred to the elevated area. Darkness was not far off but all my concerns and weariness soon faded. Quickly I realised that this slight rise, utilised for building because it was more flood-proof, contained strongly altered rock, much of it heavily stained with, or even largely composed of, iron oxides. It looked to me like gossan, the oxidised surface equivalent of sulphide minerals at depth – the very thing Geopeko was after. I took ten samples for assay from around the edges of the rise and then, under a rapidly darkening sky, retreated to the motel in West Wyalong where I would spend the night.

***

The next day I returned to Geopeko's base in Parkes, submitted the samples to be assayed for gold and copper and flew home to Sydney that night.

A few weeks later, I was back out at Parkes, getting ready to undertake another stint of mapping near Lake Cowal. Before setting off, I asked my boss if the results had come in from my last batch of samples for assay.

"Yeah, but I haven't really looked at them too closely," he replied,

"They don't seem all that exciting."

I decided to take a closer look at the assay results. Soon I found that most of the samples indeed contained little or no gold but that one of them, I think it was the seventh I had taken that evening, assayed 16g/t Au, or about half an ounce of gold per tonne. This was an excellent result in anyone's terms.

My boss was surprised but very pleased when I pointed this out. I was ecstatic – my intuition had been right. Such a result demanded follow up.

The rest of the story takes some years, and I moved on to join Pancontinental Mining later in 1980, but from that one highly anomalous sample came the discovery of the Cowal gold deposit, containing more than two million ounces of gold. The Cowal Gold Mine continues to operate today and I understand that millions more ounces of gold are thought to lie at depth. I received nothing for my efforts but the satisfaction of knowing that my diligence had created a pool of wealth that is still providing royalties to the state, taxes to the Commonwealth, value to shareholders and jobs to the local community.

That one discovery indelibly grafted the addiction of prospecting onto my DNA. Like many before me and since, it is the promise and thrill of discovery that has taken me far and wide, in all sorts of conditions and in many exotic places, seeking to find "the big one", making prospecting, perhaps more than any other, my consuming passion.

The Cowal Gold Mine in 2010

With Michael Vickers at the Hillgrove Gold Mine in northern NSW

27

THE GOLDEN MILE

EARTH

In the heart of Kalgoorlie, in Western Australia's Eastern Goldfields, there is a life-size statue of a man with a bushy beard under a high, broad-brimmed hat. He sits on a rock and in the bronze image, he is dressed in a workaday shirt, its sleeves rolled up, with drill trousers over with hard worn shoes. He is lean, wiry but clearly strong and robust. Beside him, leaning against the rock is a miner's pick and, in his hands, he holds an old-fashioned canvas water bag, from which spouts, incongruously, a modern water bubbler. He is Paddy Hannan, the man who gave birth to the Golden Mile.

Paddy Hannan discovered gold at Kalgoorlie on the 14th of June, 1893. Paddy and his fellow-Irish partners, Thomas Flanagan and Daniel Shea, were with a larger group of prospectors heading north from Coolgardie, already becoming established as a gold town in its own right, towards a rumoured new find at Mt Youle. Anxious to keep the gold find for themselves, Hannan's trio initially concealed it from the other prospectors by claiming that one of their horses had become lost overnight. In fact, they had deliberately led the horse into the bush to give them a valid excuse to stay behind the larger group, as it continued north. Once in the clear, the Irish trio staked their claim and then Hannan set off to Coolgardie to register it with the mining warden, which he did on June 17th. Within a week, over one thousand men were staking claims around the Irishmen at what became known as Kalgoorlie and one of the greatest gold booms in history had begun. It continues to this day.

Kalgoorlie (at the northern end of the Golden Mile) and its contiguous sister town of Boulder (at the southern end of the Golden Mile), hold a special place in the history of Australia. Like Broken Hill and Mt Isa, mining towns of comparable size, Kalgoorlie-Boulder sits in a remote, Outback location, in an arid environment that offers serious challenge to human habitation. Unlike those other mining towns, however, it has never really been dominated by one company and it remains today an eclectic mix of hopeful private prospectors, established gold mining companies and aspiring mineral exploration outfits, all of whom seek to emulate Hannan's historic find. This diversity creates a vibrancy about the place that surprises many first-time visitors and underpins an economy that thrives despite the vicissitudes of gold prices and stock markets. More than anything else, though, it is the lure of immense wealth from the ground that motivates people to live in such an isolated and barren setting.

My own experience of this gilded outpost started in 1980, soon after I joined Pancontinental Mining Limited ("Pancon") as a senior geologist. Pancon had begun to diversify its mineral exploration interests away from a sole focus on uranium and I had been hired to lead that diversification. Initially, we acquired a gold prospect in the Pilbara to explore. Nothing came of that prospect, other than a contact that led us to another prospector in Kalgoorlie who held leases over an old mine located about 30km north of Kalgoorlie. This mine was known as Paddington and had operated principally before World War I. It had reopened during the Great Depression of the 1930s, when the infamous rogue Claude de Bernales owned it, reputedly as a means of 'laundering' gold stolen from other people's leases.

My initial visit to Kalgoorlie was quite an eye-opener, as it is for most first-time visitors. Pancon had rented a house to accommodate its local staff, with extra room for itinerants like me. As I settled into bed on that first night the pervasiveness of the gold industry in this town was emphasised by the constant, rhythmic but muffled crunch, crunch, crunch I could hear in the distance. It was the crusher at the Mt Charlotte Gold Mine, operating just a few hundred metres away at the end of the street we were in.

When we came upon the Paddington mine, or prospect as it was then, it had been idle for nearly fifty years. The old workings were flooded, making them inaccessible from the surface, where several old shafts and mine dumps could still be seen. My colleagues and I very quickly recognised the outstanding potential of Paddington. We were especially impressed by the assay results of an old underground drill hole that had been drilled across the lode in the 1930s, probably by de Bernales to promote a new scam. That showed a possible gold orebody 25-30m wide. Pancon proceeded to acquire Paddington and over the next few years successfully delineated and then developed a major new gold deposit, amenable to open pitting. Operations commenced in 1985 and the mine quickly became a highly profitable source of revenue for the rapidly growing Pancon.

One of my favourite memories from those days is visiting Paddington when it was in full swing and being shown the gold room. That was usually off-limits to visitors but I was senior enough in the management structure to be given entry. The gold room is where the gold that is extracted in solution by cyanide leaching of the crushed and ground up ore in large tanks is recovered and smelted into bars of nearly pure gold, called doré. There is nothing quite like holding a 15kg bar of gold in your hands, which requires considerable effort because gold is heavier than almost everything else in creation. In today's values, a 15kg bar would contain nearly a million dollars worth of gold in a lump about one quarter the size of a loaf of bread. It is not hard to see why gold is so prized as a store of value. Thank goodness for colliding black holes and neutron stars, for that is where gold is born! That ring on your finger is composed of supernova ash!

The Paddington mine is now closed again, after producing over two million ounces of gold, but the processing plant continues to operate, treating gold ore brought in by truck from elsewhere in the region.

Throughout most of the 1980s I visited Kalgoorlie (from Sydney) eight or nine times a year, partly in connexion with the evaluation of Paddington and partly to supervise the company's wider exploration programme, searching for gold elsewhere in the Eastern Goldfields region. These visits, though brief (generally two or three days), were always of great interest, not least because of the gold fever that infected me as much as anyone else working in the Goldfields. Kalgoorlie lived and breathed gold! The rumour mill was very active – who has made the latest find; which district is pulling the hot money; why has WMC just staked that barren piece of moose pasture? The pubs, of which Kalgoorlie has many, were the prime source of such rumours, some with good foundation, others the result of deliberate obfuscation. One of the most popular pubs was the Palace, right in the heart of town, where Friday nights especially were boozy, boisterous dens of information and misinformation. I suspect the "Skimpies"[44] on duty behind the bar might have had something to do with it, although the Friday night "Session" had been a tradition since the nickel boom days of the late 1960s.

Another popular pub was in the sister town of Boulder. The pub

---

[44] Skimpies were barmaids dressed in see-through blouses, or less!

was called the Boulder Block and it was characterised by having an old mine shaft in the middle of the barroom floor, covered only by a metal mesh. Lights at the top pointed down into fathomless depths and who knows what stories that shaft could tell of its heyday in the original gold rush that followed Paddy Hannan's discovery? Today the Boulder Block and the ground on which it stood have gone, consumed by the ever hungry, ever growing Superpit, the largest open pit gold mine in Australia that produces around 800,000 ounces of gold a year.

Pubs were not the only outlet for social activity, either. Kalgoorlie's infamous Red Light District, comprising one block of Hay Street, was very busy at the time, or so I was told! First time visitors to Kalgoorlie were invariably driven or walked down this block after dark to see the scantily clad girls sitting in brightly lit windows, waving to passers-by and inviting them in "for a good time". Most of the brothels were little more than corrugated iron sheds but one was quite famous – Questa Casa or "The Pink House", which claims to be Australia's oldest brothel. These days it offers tours to curious visitors each afternoon. Business, I believe, is much slower than it used to be – too much competition from "sole-traders".

As exploration work progressed at Paddington it became clear that the ore body there was very similar to the Mt Charlotte deposit at the northern end of the Golden Mile, which is where Paddy Hannan made his original gold discovery. To aid our interpretation, we were able to arrange underground visits to the Mt Charlotte Mine and one or two others operating within the Golden Mile (Mt Charlotte, the Oroya, Croesus, Mt Percy, among them) to see for ourselves just what these ore bodies looked like in the third dimension.

There is nothing quite like being underground in a mine. Some mines are wet. Some mines are hot. All are dark. With no visual clues to give a sense of direction it is very easy to become disoriented, especially in a multi-level mine, where each level looks just like the one below it, or was that above? Away from active workings lighting is subdued, often absent, with the only illumination coming from the cap lamp on your hard hat. Turn that off and the darkness is complete, intense, almost a physical presence with sinister intent, muffling sound and distorting distance. To stand alongside a jumbo, its spotlights flooding the rock face and the pneumatic blast of its drills assaulting your ears is strangely reassuring. Here there are people, machines, light, life and movement. Enter disused parts of the mine, where the ventilation is disconnected, as geologists often must do, and the silence is absolute, a complicit companion of the darkness. All is still, inert, lifeless. Walk along a drive towards a distant rhythmic, sucking sound to find a pump extracting seepage from a sump beside you. Continue walking and the sound fades quickly into the consuming blackness. Step over a large rock on the floor and realise that sometime recently it fell from the roof above you. Come to the end of a blind drive and touch the rock face, as solid and impenetrable as the earth of which it's part.

But of course, you get used to it after a while. For my colleagues and me it was a familiar experience with which we felt comfortable, if not totally at ease. Not that such familiarity prevented a rather unnerving experience as we went underground at Mt Charlotte. The four of us had been in the mine for a couple of hours, trudging through endless old workings in semi-darkness, stopping from time to time to shine our cap lamps onto the gold-bearing rock where it was exposed on different levels and discussing what might be controlling the occurrence of gold. As we walked along yet another dark tunnel the mine geologist who was guiding the four of us stopped by a winze[45] that had a ladder poking up out of it.

"We'll go down here," he said, "follow me."

One by one, each of us stepped onto the ladder and started climbing down after our guide, with me in the lead. At first it felt alright; I could reach out and touch the walls around me, giving me some sense of security as I descended in near total darkness. Suddenly, however, the walls disappeared and I was surrounded by open space, very dark open space. I swung my head around, pointing my cap lamp this way and that, trying to see what was going on. The puny light failed to penetrate far into the darkness. All I could see was empty space. I looked down; the ladder continued but soon disappeared into the dark void below.

"Where the hell are you taking us?" I called to the mine geologist, my voice echoing off unseen walls.

"You'll be fine; just keep coming down," he replied nonchalantly.

I did keep descending, slowly and nervously, and eventually reached the bottom of the ladder and stepped onto a solid floor.

"Phew! That was a bit of a shock. Do you do that to all your visitors?"

"Yeah, if I can!"

Once all four of us stood on solid ground the mine geologist enlightened us.

"You've just come down through what was once the Oroya Shoot, one of the largest and probably the richest individual gold lode in the Golden Mile."

"Two million ounces of gold came out of this stope[46]," he continued as he turned on his powerful spotlight and revealed a

---

[45] A winze is a hole in the floor that usually leads down to a lower level in the mine.
[46] A stope is an underground opening in a mine from which ore is being or has been extracted.

vast, man-made cavern that seemed to go on forever. All we hopeful explorers could do was dream that maybe one day we would discover something like this within our own leases at Paddington.

Our gold exploration activities ranged far and wide from Kalgoorlie, taking us into some very remote and isolated terrain. We were well equipped, with air conditioned Landcruisers, complete with fridges, and plenty of supplies to camp out overnight. It never ceased to amaze me, however, to see just how much hard physical labour had been spent in many of these remote locations by prospectors of times past. Again and again we came upon old workings where shafts and tunnels had been hand-dug by men whose only means of transport to these isolated sites would have been on foot or at best on horseback.

"How the hell did they do it?" was my constant refrain. It really emphasised the lengths to which humans will go in the pursuit of gold.

But not only gold. Occasionally we met landholders on the vast sheep stations that are spread over the Eastern Goldfields. I remember asking one of them, a man called Fred Cock,

"How big is you place Fred?"

"Well, she's about 200 miles this way," he said, pointing south, "and about 180 miles that way," he added, pointing west. With such vast holdings, it was no wonder we did not run into owners like Fred very often.

Many of the opportunities we investigated came to us from private prospectors, who we either met in the pub (as our company Chairman had encouraged us to do) or tracked down by their reputation across town. One day my colleague, Terry Rust and I had a meeting with a prospector called Ray. Patiently, we sat around the Laminex table in his kitchen, drinking tea while he teased us with interest in what we were offering but would make no commitment. We knew he held some very promising ground up north and we wanted to negotiate an option to acquire it from him at a good price. Like most of these fellows, Ray was very cagey and difficult to pin down to a deal. Fortunately, Terry had a knack for doing this and I was happy to leave most of the negotiation to him. At one point, as the discussion continued, Ray stood up and reached up to a blue jar sitting on top of the kitchen dresser. I looked at the jar as he slowly undid its lid.

"Why is Ray choosing to apply Barrier Cream to his hands just now?" I thought.

But I had underestimated Ray. Slowly, he upended the lidless jar and spilled its contents onto the kitchen table. Gold! Nuggets and pellets of gold littered the table, the largest about the size of a golf ball, the smallest like lentil grains. Ray's display was quite calculated, as our avarice went up by several notches and it became hard to hide our eagerness.

"Jeez Ray, where the hell did you get those?"

"Oh, up the road a bit," was all the reply I got. He would not tell us where he had found the nuggets, possibly because they had come from someone else's leases (trespassing was a common enough event in the feverish 1980s gold rush days). But they did at least verify his *bona fides* as a serious prospector and the experience stiffened our resolve to do a deal, which we eventually managed to do, to everyone's satisfaction. Ray was paid handsomely in cash; we were able to explore his leases, which we did eagerly though to no avail. We never did learn where those nuggets on the kitchen table came from though.

"Ah well, there's always the next one," I rationalised. The seductive lure of the yellow metal held me firmly in its grasp and continued to do so for many years to come.

PAYDIRT

In a bleak and barren landscape of heat and flies and dust,
Old Paddy McGee, the prospector, his face a sunburned crust,
In shabby hat and faded vest, his beard like scaly rust,
Swung his pick once more to break the stony ground.
As he toiled all alone in desert isolation, no other sound
Was heard save his croaky voice as he daily would assert
"I'll be makin' me a fortune, soon's I hit paydirt."

Paddy had been a prospecting for forty years or more.
From mountain top to valley floor
The seductive lure of gold and silver ore
Had seen him trek from Kalgoorlie to Henty,
Dreaming of wealth and riches aplenty.
Heeding not his aching back and feet with blisters girt,
For 'I'll be makin' me a fortune, soon's I hit paydirt.'

Many were the times he thought his luck was in.
A promising reef of quartz that proved to be too thin,
Or a lead of gravel wash that had no gold, just tin.
Sure, there was the occasional nugget or patchy little show
That yielded up some ounces and kept him on the go.
But with real success Paddy just never seemed to flirt.
Still, "I'll be makin' me a fortune, soon's I hit paydirt."

The other diggers joked and laughed behind his back.
"Fool's gold" they'd say, their minds a single track,
"Is all he'll ever find; he'll riches always lack."

"Ah maybe they're right" said Paddy, as year gave way to year,
And failure followed failure and he heard their scornful sneer.
But still he persevered despite his critics curt,
Because *"I'll be makin' me a fortune, soon's I hit paydirt."*

But Paddy never made the hit he'd always sought.
His trusty pick grew rusty: *"Just like my mind"* he thought.
All his toil in barren soil sadly came to nought.
When finally, his eyes grew dim and he prepared to die,
"I've lived a good and healthy life" he said *"and surely that's no lie.
I've ne'er spoke ill of anyone nor any creature hurt,
So maybe that's my fortune: another kind of Paydirt."*

But, when they came to bury Paddy and say a last good-bye,
To lay him down with pious words but hearts all powder dry,
They were neither keen enough of eye
Nor sharp enough of wit
To notice, just exposed in the bottom of the pit,
A reef of quartz all white and bold
With the bright clear glint and sparkle of true metallic gold!

Today they say if you come and stand beside that grave
And strain your ears, bend your back and, well, be a little brave,
You will hear a gentle tapping within the hallowed cave.
A hammering and scraping and other digging sound,
And above the ghostly clatter in words at last profound,
If you listen very carefully and are especially alert,
You'll hear: *"I'm makin' me a fortune, now I've hit paydirt."*

28

NAKED HEIKKI

FIRE

On a bookshelf in my study there is a certificate that I treasure greatly as a memento of my working life. One glance evokes fond memories of a time when I was forced to abandon restraint and embrace a culture that, like the place itself, is about as far away from Australia as you can get, and still be on earth! Here is a copy:

SAUNA CERTIFICATE

THIS IS TO CERTIFY THAT
OUR HONOURED GUEST

MR GREBY G. LOUGHED

HAS TODAY SHOWN EXTRAORDINARY STRENGTH
OF CHARACTER (SISU) AND SURVIVED ALL THE
ENJOYMENT A RED-HOT FINNISH SMOKE SAUNA
CAN OFFER

THE TEMPERATURE MEASURED
DURING THE TIME OF ORDEAL

94.2 °C
2015 °F

8 TH APRIL 19 83

OUTOKUMPU Oy
PYHÄSALMI MINE

MASTER OF CEREMONIES

The Finns are an independent and innovative race of people but some of their customs take a bit of getting used to. Where else would you find yourself naked, in a sauna, at 94 degrees Celsius, with people you have just met? The sauna is central to life in Finland; babies are born in them (though not while they are hot of course) and foreigners die in them, well almost, but no visit to Finland is complete without a genuine sauna experience.

My exposure to this peculiar culture came about in the mid 1980's, during a series of business visits to Finland for meetings with a company called Outokumpu Oy. Outokumpu is the largest mining company in Finland, a bit comparable with Australia's BHP. In fact, "Outokumpu" is Finnish for "strange-looking hill", not too different from "broken hill". In my business development role with Pancontinental Mining ("Pancon"; my employer at the time) I had introduced the much larger Outokumpu to Pancon as a potential joint venture partner, with the Finnish company later becoming our actual partner in the Thalanga lead-zinc mine in Queensland and joining us to prospect for base metals across Australia.

My initial visit to Finland was made as part of setting up the relationship, where my Finnish host was Heikki Solin, assisted by his Chief Geologist, Heikki Wennervirta. Our first meeting took place one March afternoon in Mr Solin's Helsinki office. After a couple of hours of formal business discussion, I was invited to join my hosts for a sauna. I had been forewarned and was kind of expecting it but that did nothing to prepare me for the experience I was about to undergo.

As a group, we went down in a lift to the building's basement and entered a bright and clean change room. There suits, ties and shirts were shed, then all other clothing and I was given a small cloth towel. My hosts then moved through a doorway into a darkened room. I followed, more than a little self-conscious. It was already hot in the sauna and anxiously I watched my companions to see what I should do. The little towel, it turned out, was to sit on, not for modesty. As I sat, naked, in the sauna, with professional businessmen I had just met, the first problem was to know where to look. But, as I soon discovered, there is a well established sauna protocol, and the darkened room was merciful to me, until my eyes adjusted. Before long, I began to relax and just blend in, which is surprisingly easy when you are all naked.

After some time, we returned to the change room and showered, before squatting, towels strategically draped, on benches to consume a couple of beers. Then it was back into the sauna, hot as ever, for another session. The conversation continued, as business-like as if we were sitting around a board room table. After perhaps an hour all up we retreated again to the change room, showered and dressed back into our suits and ties. Minutes later, we left the office building to go to dinner. By then it was snowing and I was struck by just how far away from Australia I was, both geographically and culturally.

The next day, I began a tour of some of Outokumpu's operations in various parts of Finland, accompanied by Heikki Wennervirta. Heikki was a tall, striking man, with the profile of an Easter Island statue and a cutting sense of humour. He was a superb storyteller, with a seemingly endless list of tales and anecdotes, many of which were merciless to the neighbouring Russians. His favourite toast was:

"Here's to the Russian navy! Bottoms up!"

But Heikki's most remarkable attribute was his voice, which had a deep resonant timbre that sounded like a slow, rolling timpani solo as he articulated every syllable.

Most of what I know about the history of Finland, a proud but often beleaguered nation, I learned from Heikki on that trip. His combination of humour and gravitas, delivered in his deep base voice, was captivating, even at ninety degrees. The instructions began at our first stop, the Outokumpu plant at Kokkola, 500km north of the capital. After a tour of the operations, we were conducted to the company's guest lodge, situated in an isolated part of the property, where a snowy landscape merged with a frozen Gulf of Bothnia. Everything was white; the bumpy bit in the foreground was land; the flat expanse beyond was the sea. No sooner had we found our respective rooms and returned to the lounge than Heikki's deep voice boomed:

"And now we go to sauna!"

He pronounced it: "saa-oo-naa".

At least I knew what to expect and dutifully trotted off after Heikki, with the plant's manager and his assistant, to a large sauna installed in one corner of the lodge. Once there I was regaled with the complete history of Finland, recited in remarkable detail by Heikki. At first, I was entranced; how could I not be listening to that wonderful voice? But as the temperature rose and the sweat began to pour off my nose and dribble down my chest, my concentration began to wane. Finally, by the time Heikki got to the 1850s, I declared:

"I'm afraid I can't take another hundred years in this sauna. Can we have a break?"

"Of course," said my host, as we exited the sauna and sat in the change room sipping our beers.

But I knew it was only a temporary reprieve. All too soon, we were back in the sauna and the storytelling resumed. I listened as attentively as I could, almost, almost, getting used to the heat. Eventually, the history recital came to an end and the conversation turned to other things.

"I thought you Finns went and rolled in the snow after your sauna," I remarked, trying to lighten up a bit.

"After you", said Heikki, as he leaned over to a door and opened it, revealing a cold, snowy vista. I declined the invitation, which was wise in view of his next story.

Some previous guests at the lodge had indeed left through this door to roll naked in the snow. Unfortunately, or as Heikki would say, "un-fort-u-nate-ly", the last man out forgot to unlock the door as he left. Outside it was -20°C and dark and there was no one else in the lodge. They ran around to the front door but it, too, was locked. They were in serious trouble. One man volunteered to run to the gatekeeper's house, some 2km away, for help. He reached the house and knocked desperately on the door. It was opened by the gatekeeper's daughter, who, seeing a naked man standing there in the dark, screamed and slammed the door. Meanwhile, back at the lodge, the others managed to pry the handle of the front door off and use it to break a window and enter the lodge. All ended happily, but afterwards the plant manager vowed wryly: "Next time I will get a stronger door handle".

Several more visits to Finland ensued over the next few years, as the relationship between Outokumpu and Pancon was established and blossomed. Most often I seemed to find myself there in late winter or early spring, I suspect because I quite enjoyed the climate contrast with home. On one visit, we went to the city of Outokumpu, where my host company had its origins with a large copper-nickel mine. The guest lodge sauna on that occasion was a wooden hut outdoors, with a log fire for heat, instead of the usual electric radiator. Here there were also birch branches for the traditional back slapping ("brings the blood to the surface" I was told). Outside the hut it was a balmy -13°C at the time (no wind and weak sun made it just tolerable).

One of the most memorable visits, made in the company of Pancon colleagues Angus Collins and Paul Wilson, was to Rovaniemi, on the Arctic Circle in Lapland – the Finnish "Outback". There, after meeting with our Finnish Geological Survey hosts, we were invited to try to emulate their winter field excursions. This involved donning orange boiler suits and blue beanies, slinging a backpack over the shoulder and a geological hammer into the belt pouch, putting on a set of cross-country skis and then attempting to traverse the countryside to look at rock outcrops. Rather different from Australian Outback field work! The Aussie trio was out of its depth, in more ways than one, but we did enjoy the friendly barbecue in the snow afterwards. Later we were taken to see "Santa's House", parked right on the Arctic Circle and complete with live reindeer in the front yard (none with red noses though). A walk across the Kemijoki River, secure in the knowledge that there was more than a metre of ice beneath our feet, was another treat.

At each mine site or exploration base camp the ritual was the same: Hours spent roasting in a sauna, followed by boozy dinners with our Finnish hosts[47], whose capacity for alcohol absorption is astounding. We tried to keep up with them by drinking *Lapin Kulta*[48], the local beer, while singing folk songs, alternately Finnish and Australian, until late in the night. It was a hopeless cause; once we had delivered "Click go the Shears", "Walzing Matilda" and a measly few others we were stumped. The Finns continued undaunted long after we Aussies had slunk off to bed. Only later did we discover that one of our hosts, Markku Mäkela, who was the image of Omar Sharif, had paid his way through university by singing in nightclubs.

Great as those sights and those memories are, for me the abiding

---

[47] Reindeer steak was often on the menu and sometimes even bear meatballs.
[48] Finnish for Lap Gold.

image of Finland is auditory – I shall never forget the sound of Heikki Wennervirta's voice.

Even now, when I recall his command …

"And now we go to sa-u-na" …

my skin tingles with a mixture of anticipation and fear.

With colleagues Angus Collins and Paul Wilson, standing on the frozen Kemijoki River in Rovaniemi

I try my hand at winter field work, Finnish-style

With our Finnish hosts on the Arctic Circle in Rovaniemi

Early morning at the lodge in Outukumpu - a balmy -13 °C

29

RAINFOREST REALTOR

EARTH

If you fly just 100km directly north of Port Moresby in Papua New Guinea you will find yourself amongst steep and rugged mountains that are shrouded in cloud for most of each day. When the cloud clears, you will see a landscape draped in the dense rain forest that covers most of this vividly green country. It was here in this stunningly beautiful but logistically challenging environment that Newmont Mining discovered the Tolukuma gold deposit in 1985. Initially there was great excitement in the Newmont ranks as further exploration revealed very high grades of gold in the mineralised rocks. The company proceeded to drill out the discovery over the next few years – a very costly exercise given the need for helicopter support throughout. As they did so, the high grades were confirmed as continuing to depth. However, Newmont's enthusiasm for their discovery waned as it was realised that, rich as it was, the total size of the gold resource was quite limited. In the end, it proved to be too small to meet the investment criteria of a large company like Newmont. So, in 1992, a decision was made to divest the property – to sell it to a smaller company for whom the deposit would be material indeed.

Responsibility for taking this action fell to my old mate from Kayubulan Ridge, John Dow, who, at that time, held a very senior position with Newmont and was based in Jakarta.

"We need someone with both geological knowledge and an ability to promote the financial potential of the deposit," John told me in a phone call from Jakarta, "not just some overpaid and over-confident merchant banking type who'll charge like a wounded bull."

"I think you are the bloke to do it," he concluded.

John's confidence in me was flattering but a little unnerving – I had never done anything quite like this before.

"It'll be a challenge, but I think I'm up to it," I told him, "I'd like to give it a go. No guarantees mind you."

"Not required," was John's encouraging reply.

So began one of the most challenging but financially rewarding experiences of my working life. At John's instruction, I was given an exclusive mandate to carry out the total divestment of Tolukuma for Newmont. I would be the dog's body on the ground and be responsible for digging up the potential purchasers from across the Australian mining fraternity. Assisted by a senior solicitor from Newmont's law firm in Sydney, who would look after the legal documentation, I would also have to negotiate the sale terms with the successful bidder on Newmont's behalf. My engagement with Newmont provided attractive remuneration on a *per diem* basis for the time I put into the task. Best of all, though, I would be entitled to a healthy commission on any part of the sale price above a base level of $US5 million. Material as that was to me, it was but a fraction of what a traditional merchant bank would charge Newmont for effecting this transaction.

The divestment process began with the compilation of all existing geological data concerning Tolukuma into "data room" that I rented in a serviced office complex in suburban Lindfield. Representatives of prospective buyers first visited the data room to review the numerous reports and maps from Newmont that I had assembled, detailing what had been done already at Tolukuma. From this they had to form a view on the project's potential and whether a site visit was justified. For those keen enough to continue, a visit to the prospect was then arranged. This necessitated my making multiple trips to PNG, setting myself up in the Tolukuma camp and hosting my visitors on successive days.

Access to Tolukuma was by helicopter only and most flights took place in the morning. Each time a prospective buyer's representative (a geologist) came to see the place I either flew in with them or was there to greet them, standing anxiously at the edge of a tiny helipad cut into the ridge just above the camp, like a nervous realtor before his first "open inspection". Tolukuma was situated on a sharp ridge, about 1550m above sea level, with a spectacular view across a valley deeply incised into the mountains. The intense green of the rainforest contrasted with the paler green of the kunai grass that capped many of the ridges. Communication with Port Moresby, just 100km away, was by radio-telephone only, with reception poor at the best of times and non-existent when the daily rain storms came sweeping up the valley. The mornings were the best time of day, with the sky relatively clear and the air still. Here and there smoke from fires in the villages dotted around the ridges and spurs of the valley below could be seen. At those times, it was possible to hear village people calling to each other across the void, their voices wispy and ethereal, like the smoke of their fires, drifting up into the tranquil air. By midday the clouds had usually descended, closing us in and smothering both sight and sound. Any form of airborne activity was severely limited in such conditions.

There was no road to the site at that time and even when the successful purchaser eventually developed the mine, no road was built. All equipment was flown in by helicopter, including the use of a giant Russian aircraft with a payload of 15 tonnes. There was an airstrip across the valley from Tolukuma, just ten minutes flying time by chopper but a day away by foot. That strip was typical of small landing fields in the PNG Highlands. The crest of a spur ridge had been shaved off, creating a grass landing strip that was short but so steep that when a fixed wing aircraft landed, always uphill, it would stop almost instantly. The pilot then had to apply substantial power just to get the plane up the slope to the small horizontal parking area at the top end of the runway. Taking off was exhilarating – power on, run down the slope and airborne in seconds. Just as well, as the land dropped away rapidly at the bottom end of the strip.

As my prospective buyers came, one by one, I adopted the best tradition of your local real estate agent, extolling the virtues and glossing over the difficulties. "Open inspections" took at least a whole day and involved trudging up and down a criss-crossing series of tracks that covered the prospect and exposed the high grade gold reefs on steep cutaway banks. Some of the tracks were quite hazardous, barely half a metre wide, with a cliff on one side and a precipitous drop off on the other. To make matters worse, it was often raining as we clambered around the mountain. Even when the rain held off, the ground was saturated and the tracks were slippery obstacle courses. I was very fit and took vicarious pleasure in setting a brisk pace as we climbed around the mountain. In some places the local labourers had built steps by laying pieces of wood across the pathway. These helped, except when they gave way due to rapidly advancing rot and caused my visitors to slide painfully down the steep slopes till they hit something solid, usually a tree.

"Ah, yeah. I meant to tell you not to trust the steps too much," I apologised, tongue in cheek, after such events.

"It's been a while since Newmont did any track maintenance around here."

Fortunately, I lost no potential customers to the local hazards and was able to put them on the helicopter to return to Port Moresby a little bruised but in one piece the next morning.

Most prospective buyers who came to Tolukuma liked its geological potential, particularly the high grades of gold discovered in Newmont's drilling (20-30g/t Au, or nearly one ounce of gold per tonne), but were daunted by its mountainous, inaccessible location. I was secretly sympathetic but played down those difficulties, as a good real estate agent should.

"Remember what the three best attributes of a successful mine are," I would say, trying not to sound glib, "grade, grade and grade. We've certainly got those here." That much was agreed but I was glad they did not then point out that Tolukuma lacked the three most important attributes of good real estate: "Position, position, position!"

It took several sojourns at Tolukuma and much solid talking on my part but eventually I was successful in finding a buyer who was prepared to pay a price that Newmont would accept. That buyer was a small Australian company called Dome Resources and they subsequently went ahead and developed the mine. Later they were taken over by a South African group.

Fortunately for me, the agreed sale price was $US9 million, comprising $US5 million up front, with a further $US1 million a year later and the final $US3 million in staged payments several years later, when certain production targets had been met. Those obligations carried over with any change of ownership on the part of the purchaser. I still had to chase the final three commission payments but, with Newmont's help, they were eventually paid. Those sums not only consolidated my solvency at the time but gave me the capital to start my own exploration company four years later.

30

THE DARK SIDE

Some of my former colleagues in industry considered that I had "deserted the ship" when I took up the government role as Director General of the Department of Mineral Resources ("DMR") in NSW. My response to their consternation was to jokingly claim:

"Yes, I've gone over to The Dark Side of the Force!"

Star Wars was very topical in those days and this epithet produced a few sniggers. I think a lot of them saw it more as an expression of stark reality, such was industry's jaundiced view of the DMR at the time. I was determined to change that view.

My appointment as head of the DMR meant that I changed roles from seeking to create wealth by discovery of mineral resources to facilitating and enhancing the ability of others to do so, legally and responsibly. In some ways, the DMR position was a "homecoming", as I had started my geological career nearly thirty years beforehand (prior to going to Berkeley) as a geologist with the NSW Geological Survey, a major division of the Department I now headed. In 1965, it was called the Department of Mines and as a very junior geologist, little did I realise then that I would return decades later as the chief.

A large part of my charter when appointed to the DMR in 1993 was to revitalise a moribund bureaucracy by introducing modern management practices, referencing my time at Harvard and experience in the private sector. To achieve this meant giving the employees, many of whom had been there all their working lives, a better understanding of their place in the world and a broader perspective on the reason for the DMR's existence. Many in the Department, however, thought I was only there to decimate the numbers.

"He's just a bloody axe man," the rumours said.

Wrong!

"He's not really one of us."

That much was true at least!

Against such headwinds, it took some time to build trust and cooperation. Slowly, though, as I introduced change without sacking anyone, their faith in me grew.

One of my main concerns was a lack of teamwork and the prevalence of fiefdoms, with the various branches and divisions jealously guarding their territory. There was no real sense of a body corporate and common objective. The Department as a whole was charged with custodianship of the mineral resource endowment of NSW, a major public asset. This involved regulation of the industry exploring for and mining those resources, including exploration monitoring and support, environmental protection, mine safety, titles administration and royalty collection. We were also required to promote NSW as a mineral resource investment destination, in competition with the other Australian states and foreign jurisdictions. Each group knew its role but saw it as an end in itself, rather than as part of a whole enterprise. My first task was to break down those barriers and give my staff a sense of comradeship and a new vision of their role in government.

To catalyse this process, I decided to try something I had seen work well in industry – an out-of-the-office management retreat. This I saw as consistent with my new management practices charter. With help from a few key, progressive staff, I put together a three-day series of training seminars, discussion groups and other communal activities for the fifty most senior people in the Department. To undertake this exercise, we retreated to a large but modest motel, with conference facilities, at Leura in the Blue Mountains and began the experience. The early signs were good, as people from the various divisions started to work together and accept that everyone had a worthwhile role to play in achieving the Department's goals (which we redefined in the process). By interacting on neutral territory an understanding and appreciation of each other's place in the organisation began to emerge.

I was much encouraged by the positive comments that came from the participants as we went along, feeling that these validated what I was doing. But I had underestimated the pettiness of traditional bureaucracy. Someone in the Department, probably the one ranked number 51, decided to attack my initiative by calling a very prominent talk-back radio host and claiming, quite falsely (we were not at the *Fairmont Hotel* after all), that we Government employees were swanning it in luxury at public expense. The "shock jock", without checking the facts, was highly critical of me and my Minister for allowing such a waste of public money.

"If they don't know how to manage, what are they doing as managers?" he pompously announced on air.

When told about it I was both angry and upset. The Minister was livid. The negative publicity, though totally ill-informed and incorrect, was most unwelcome. In a strained telephone call to the Minister I meekly tried to explain my motivation and purpose and explain that I saw it as entirely consistent with the "modern management practices" charter he had given me. The whole affair was a salutary experience for me. As someone with no political background, it served as a harsh lesson that in government, politics, rightly or wrongly, outranks all other factors. Even so, at the end of the retreat I was greatly mollified by the upbeat sentiment expressed by the wide range of staff who attended. It was particularly gratifying when a senior person in the Geological Survey came up to me and said:

"I've worked in this Department for 23 years and today I met someone for the first time who has been here just as long."

One of the most troublesome parts of my Director General duties was the regulation of the opal mining industry at Lightning Ridge. Opal miners operated on small claims and most were one-man shows. The tough, secretive men (and a few women) guarded their plots jealously and were suspicious of any outsiders, especially government representatives. Few, if any, admitted to finding opal but the city-based merchants who came to buy from the miners always seemed to leave town with their briefcases full. An aversion to paying tax probably had something to do with the miners' regular lament of "dry diggings".

Many miners lived right on their claims, building illegal houses (referred to as "Camps on Claims") some made from local stone or even empty beer bottles:

"All the better to guard it from thieving bastards!" was their justification.

Safety was my main concern, as few of the miners had formal mining qualifications and far too many were of the "she'll be right" school. The underground workings were typically about 15m below surface and were accessed by narrow shafts drilled through the soft overlying rock. Roof collapses underground were common and robbing pillars that showed flashes of opal was a common mistake. The pillars were supposed to support the roof and shaving rock from their sides ultimately caused them to fail. More than one miner, too, has died from carbon monoxide poisoning caused by using petrol-driven generators underground without adequate ventilation. Years later, I was inspired to write a simple bush ballad, aimed at capturing the spirit of Lightning Ridge, surely one of Australia's most engaging but enigmatic Outback towns.

Before taking up the position as Director General I had flown to Adelaide to consult an acquaintance who had found himself in a similar position two years ahead of me. Ross Fardon was a well know exploration geologist with a PhD from Harvard and a long history of success with major Australian companies, including Western Mining Corporation (WMC was the best explorer of them all) and BHP, as well as smaller, more entrepreneurial outfits. I quizzed Ross about the adjustments necessary for the transition from industry to government employment. He identified several issues I had not thought of, but none fatal, and strongly encouraged me to give it a go. He was keen, I think, to see another ex-industry person as a colleague in the public sector. Ross was already well advanced in promoting the exploration potential of his state, up till then seen by industry largely as a barren wasteland, even though the giant Olympic Dam deposit had already been discovered there (by WMC; who else?). At that meeting, I proposed to Ross that, if I went ahead, we should bring our respective Geological Survey divisions together at Broken Hill to update and extend our knowledge of the geology of the region jointly. Broken Hill geology is extremely complex and resolving the long geological history of this region was crucial to aiding exploration for new deposits of similar kind. The great mines of Broken Hill have been of immense value to both states; to NSW for the vast royalties and economic activity it had generated for over a century and to SA for the smelting and refining industry at Port Pirie it had supplied for just about as long. It behoved us both to encourage a major scientific re-analysis of all that had been done before us.

The Broken Hill Lode lies within NSW but only about 40km from the South Australian border. It is the largest silver-lead-zinc deposit in the world but, unlike most other large ore bodies elsewhere, the Broken Hill deposit seems to be an only child. The next biggest deposit known in the geological province in which it lies is orders of magnitude smaller. For decades, this had been seen by geologists as a glaring anomaly and explorers had been searching vainly for the siblings of the Broken Hill Lode. The geological province that hosts Broken Hill (called the Curnamona Province) is split roughly 30/70 between NSW and SA. Up till then, government and most industry geological investigations had also been split along that state boundary. My idea was that we should pool our knowledge and base our studies on geological boundaries, not political boundaries. Ross Fardon heartily agreed and the Broken Hill Exploration Initiative was born; it subsequently became a long-standing entity, strongly supported by the mining industry and the Federal Government. Sadly, we are still to find that elusive sibling.

Fardon's most impressive achievement in his role as Head of Mines and Energy South Australia had been convincing his government to fund a new initiative that required them to spend $15 million on the acquisition of new, high quality airborne geophysical data. This was a bold move, given that South Australia was almost broke at the time, following the State Bank fiasco. The new information was published and provided to industry at low cost as a lure to spark their interest in the moribund state. Industry had responded enthusiastically and a new wave of minerals investment was sweeping into South Australia.

As I watched from Sydney, I became determined to emulate that feat in my own state. After careful preparation, with valuable input from my Departmental staff, especially from Margaret Campbell, a key colleague who was hard-working, perceptive and an ardent supporter of my attempts to re-invigorate the Department. Together, we prepared the case to go to Cabinet. In the end, we convinced the NSW Coalition Government of the day to allocate to my Department $40 million of additional expenditure over five years, up to the year 2000. This money was committed to the generation of new, pre-competitive geoscientific data and the means by which to disseminate it to industry investors in digital form. I called it *"Discovery 2000"*, an allusion to *"Sydney 2000"*, the very topical name and logo used for the impending Olympic Games in Sydney. I also wanted to emphasise that this initiative was about outcomes ("discovery") and not just processes ("exploration") – something of a novel concept for many government employees.

My request for funding progressed slowly through the standard public service channels. Following advice from Percy Allan AM, then Head of Treasury, I had submitted the application as a Capital Request, not as Recurrent Expenditure. This proved crucial and eventually I was called to a meeting of the Cabinet Expenditure Review Committee to explain to the Premier, John Fahey, the Treasurer, Peter Collins and the other ministers present what we were proposing and why they should fund it. I took along some props, including an aeromagnetic map of the Bathurst 1:250,000 map sheet area[49], recently published by my Department. This map collated the results of recent airborne magnetic surveys of the Bathurst region in the form of a brightly coloured collage of blue, green, yellow, orange, red and purple. The warmer colours (oranges, reds and purples) showed the more highly magnetic areas, while the blues and greens were less magnetic. It was a kind of mega-scale MRI of the region. This information allowed geologists and geophysicists to interpret sub-surface geology and especially structure, even where there was no rock exposure at surface.

---

[49] It was one of the few regions to have such coverage in NSW at the time.

"It looks like the Great Barrier Reef," exclaimed the Premier, as I laid the map on the table in front of him.

"Yes it does, Premier."

I tried not to sound too obsequious!

"What we need to do is cover all of NSW with this sort of mapping as an incentive to make it easier for companies to select areas in which to explore," I continued, adding that I had often used government maps like this as a guide while working as an exploration geologist in industry.

"When I was working for Geopeko out at Parkes in the 1970s we used this kind of information, primitive as it was then, to select the area north of West Wyalong where we later found the Cowal Gold Deposit. I actually made the first gold discovery there."

That non-bureaucratic background and record of exploration success gave me a credibility that few of my predecessors could have claimed.

"South Australia is way ahead of us in generating this kind of thing thanks to extra funding that's been provided by the SA Government for their Exploration Initiative. Mining companies are flocking to that state to take advantage of their new high quality database.

They're investing millions in high-tech mineral exploration. South Australia is suddenly the place to be for minerals and we're missing out."

The Premier's eyebrows were furrowing and I could see his interest growing, so I swallowed, breathed deeply and continued boldly.

"And yet, on geological fundamentals, NSW is more prospective than South Australia. We just can't offer our investors the same head start."

Then I went in for the killer punch.

"We really need to get cracking if we don't want to be outdone by South Australia."

That clinched it.

"Well we're not going to be beaten by South Australia," Mr Fahey declared.

"You can have your money."

"Now get out of here," said Committee member Bruce Baird, Minister for Transport, with a twinkle in his eye. That twinkle reflected the fact that Bruce was an old friend of mine, going back to the 1960s, when, among other things, he was best man at my wedding.

As soon as I got back into my car I picked up the big, chunky Carphone installed under the dashboard and called Margaret Campbell in my Department.

"We're on!" I declared.

Margaret was ecstatic.

When I returned to the office in St Leonards there were grins everywhere. This had never been done before. The extra money would re-invigorate the Geological Survey especially but its effects would also be felt right throughout the Department. Morale took a giant leap upwards and my staff began to think:

"Maybe this interloper from industry is not so bad after all!"

Discovery 2000 was launched immediately and was instantly applauded by the mining industry. The DMR was no longer a pariah.

A year or so later a Labor Government was elected and I had to sell the concept to them anew. To their credit, they saw the value of what we were doing and continued the funding, although reducing it to $35 million over the five-year period. Fortunately, I had had the foresight to stagger the expenditure, with $12 million in the first year, $8 million in the second year and lower annual amounts for three years after that. By the time the Carr Government was elected the money was nearly half spent.

During its tenure, Discovery 2000 saw renewed recognition by industry of the resource potential of NSW. Within two years, annual mineral exploration investment in the state more than doubled to over $100 million. Similar initiatives, by various names, have continued in NSW for much of the post-2000 period, leading to continuing investment in the State's resources sector, generating discoveries, jobs, production, taxes and royalties for the owners of those resources – the citizens of NSW.

For someone who had worked so long in the private sector, being a public service "mandarin" was a challenging, at times frustrating, but overall rewarding experience. There were some extremely competent people in my Department, dedicated to doing their jobs efficiently and well, providing a professional service to the people of NSW that, while not highly visible to the public, was essential to the continued prosperity of the state. The Mines Inspectors were vigilant in maintaining safety standards in NSW coal mines, metal mines and quarries. The Titles staff managed the complex administration of legal titles to minerals across the state[50] with fervour and the administrators diligently collected the royalties from operating mines that made a major contribution to the State's coffers. Meanwhile the Geological Survey continued to map the geology and geophysics of NSW in ever greater detail. Their products included maps, books and other publications, with a strong emphasis on quality and accessibility. Chief among the new products were the superb aeromagnetic maps, in paper and digital format, made possible by Discovery 2000.

---

[50] Mining leases, exploration licences, mineral claims and assorted other titles.

A prime information resource for would-be explorers was and still is the thousands of reports on work carried out in exploration licence areas by the licence holders. The Mining Act requires that these reports be submitted to the DMR every six months to be held as a permanent record of what work was done and how much money was spent[51]. The reports are confidential while a licence remains valid but revert to open file once the licence over a given area is dropped or expires. They are then made available to anyone researching the area for possible re-examination and renewed exploration. This research is a tedious but fundamentally important part of the exploration and discovery process. Many a new deposit has been found on ground previously examined by others[52]. New geological interpretations, new technology, or simply greater diligence are the reasons this is so. Thanks to Discovery 2000, there was funding to put these hard copy reports, prone to deterioration and theft, into digital form that would allow secure access via the Internet, which was just then emerging. Soon "DIGS", as it was called, was seen as a world-beating innovation and was copied in other Australian states and in Canada.

Discovery 2000 gave the Geological Survey staff renewed enthusiasm and self-respect and my commitment to supporting their high quality earth science was much appreciated. When I left the Director General role in 1997 to return to industry one of the research scientists in the Geological Survey wrote a letter to me that I treasure greatly.

"I have worked in this Department for 27 years," he said, "and

---

[51] Each Exploration Licence has a minimum annual expenditure and the Geological Survey assesses the reports submitted to confirm that the results reported match the expenditure claimed.
[52] The "Century" lead-zinc deposit in far North West Queensland was so-called because the area had been prospected on and off for 100 years before the discovery was finally made.

the last four years while you have been here as Director General have been by far the best."

REDFIRE

A stranger appeared suddenly in Lightning Ridge,
Conspicuously out of place, like a boat on a bridge.
The locals kept their distance but cast furtive glances,
Uncertain, suspicious, taking no chances.
Who was this man in coat and tie, well dressed?
A cop? A thief? Or even worse they guessed,
A tax collector here at government behest.

Now, mining for opals is not everyone's cup of tea.
The work's too hard and dirty for the likes of you and me.
But for some hardy souls it becomes a kind of habit,
Excavating underground like a giant demented rabbit.
Dodging falls of rock from ground always unstable,
Tripping over tools and coils of electric cable,
While barely managing daily to put food upon the table.

It is the lure of precious opal that drives these men to toil
In semi-darkness, all alone, beneath the desert soil.
Opal, a gorgeous mineral with colours rich and rare,

Whose distribution underground is patently unfair.
One miner labours vainly for many a long hot stint,
While another, just beginning, of colour spies a hint
And unearths a precious stone with dazzling spectral tint.

But even when a lucky digger finally strikes a run
Of brilliant solid opal more fiery than the sun,
He must resist the urge to jump and shout for joy,
Developing instead quite a clever little ploy
To conceal his hard won precious find,
And guard against those rogues unkind
Who would tax away his profit, leaving none behind.

Looking like someone down on his luck
He'll continue driving his clapped-out truck;
Hitching up his pants with a length of old rope;
Armpits never making acquaintance with a bar of soap,
For all the world a sorry sight,
A man clearly in the most desperate plight,
Yet quietly working his claim by day and by night.

Yes, the Ridge is a place of eccentric old characters,
Driving around in their trucks and big tractors,
Living in houses made of stone or beer bottles,
And revving their engines with manual throttles.
Claiming as always to have made no discovery

And be waiting for fate to make a recovery,
While building a fortune through quiet skulduggery.

So, don't be concerned when you come to this town
And are greeted with fear and a sceptical frown.
Don't be surprised if their names are all Smith
And the presence of opal is said to be myth.
Just be sure when you're there they know all the facts
Of your good intentions and charitable acts,
And never ever mention or talk about tax!

---

5. CONTINUING PASSIONS

31

A LIFE CYCLES

AETHER

From rainforests to the Outback, from mountains of fire to the meaning of life, geology is for me the most engaging and absorbing of the natural sciences, one that continues to enrich and empower my life, even as that life cycles and I decelerate into retirement. I am still enthralled by rain forests, I still adore the Outback, I am still in awe of volcanoes and most of all, my fascination with the natural world still validates my very existence.

Retirement has given me the time to pursue these interests and to explore and reflect on what has been of value in my life. Of course, family, friends, religion, art, music, wealth and a lot more of the joys of life all have their place, but not in this volume. Here I examine only the extraordinary gift that knowledge and love of Planet Earth has been for me. My passion for the Earth has provided much to be thankful for, many memories that are treasured and countless experiences that leave me now feeling that I lack for nothing. My life has been full and fulfilled and if that amounts to meaning in life then I have achieved it and am satisfied with my place in the cosmos. I seek nothing more.

Now, once again like a diligent geologist, we will sample a few of these my Continuing Passions.

32

RAINFORESTS

"One thing is for sure: When I graduate and start working as a geologist I'm going to stay well away from rainforests."

Consoled by this thought I laboured on, swinging my machete left and right as I hacked my way through the vines and creepers, the ferns and nettles, the shrubs and briars of the Minnamurra rainforest. It was 1964 and I was struggling to meet the challenges of the area assigned to me for study in my BSc Honours year in geology at Sydney University. It was not a large area, roughly 40km^2 covering the Illawarra escarpment and adjacent foothills just west of Jamberoo, near Kiama on the NSW south coast. But it was steep. At least the escarpment was, all 2,000ft (650m) of its vertical extent. And it was covered in temperate rainforest, watered regularly by the humid coastal breezes that wafted in from the Tasman Sea and swept up the escarpment, dropping their moisture as they went.

The focal point for my geological investigation was Minnamurra Falls Reserve[53], a small nature reserve operated by the local council and managed by the inimitable Howard Judd. Howard was an extraordinary man, endlessly inquisitive, friendly but with an air of authority. The council employed him as ranger to manage the reserve and educate its visitors, especially schoolchildren, a role he filled with alacrity. He was a tall man with a kindly face framed by puffy, suntanned cheeks stitched together along deeply etched wrinkles. A

---

[53] Now known as the Minnamurra Rainforest and part of Morton National Park

self-taught botanist, he applied himself to studying the diverse plant life that grew within the reserve and nearby on the escarpment. There seemed to be not a single local plant that he could not identify and assign to its rightful botanical classification. When I knew him his one outstanding quest for life, as he often told me, was to visit the rainforests of the Amazon, a goal he achieved some years later.

Howard was an avid naturalist but he knew very little about rocks so when I turned up to study the geology of the area he greeted me with enthusiasm and offered hospitality that greatly facilitated my work in the area. He was a gifted story teller, ever keen to communicate the natural history of his little kingdom to visitors and able to entertain young and old alike with tales from the rainforest. His instinctive curiosity led him to question me incessantly about the rocks I was finding. Eventually he added geology to the spectrum of information he imparted to visiting school children and members of the public, who dropped in to the reserve for a day's picnic and a walk to the falls.

"Where did you get to today?" he would ask in a broad drawl that belied an underlying sharp mind. It became something of a ritual when I returned to the little cottage beside the Minnamurra River after a day in the bush. Howard had made the cottage available to a student colleague, Al Raam[54], and me as a place to stay while we worked in the district. Located at the entrance to the Reserve it was an ancient little farm house, four basic rooms and a veranda that looked out through a giant camellia tree – "the biggest in the Southern Hemisphere" claimed Howard – to lush green paddocks, replete with mooing dairy cattle. The cottage and the Reserve were reached by driving over a causeway across the river, usually no more than a brisk creek, but after rain, a raging torrent that more than once kept Al Raam and me at home.

---

[54] Al was studying the volcanic sequence on the coast at Kiama.

Howard's quest to identify the full range of Minnamurra plant life had taken him through much of the adjoining forest, well beyond the tracks and paths used by visitors to the Reserve and into the domain that I was then mapping and sampling. He knew what I was talking about when I told him what I had seen in rock exposures in waterfalls higher up the escarpment, well above the zone seen by visitors. Places where igneous rocks had intruded beneath coal seams, turning them to coke, or where ancient lava flows displayed the distinctive pillow structures that characterise submarine eruption.

I respected Howard greatly and very much enjoyed our chats on botanical and geological topics. Much of the deep affection I now hold for the rainforest can be attributed to him and the reverence he imbued in me for the vibrancy, diversity and resilience of its life, both animal and vegetable. In return, I was able to awaken his mind to the importance of the underlying rocks in shaping the landscape and generating the fertile soils that sustain all that vitality.

But at the time, it was just bloody hard work, scrambling through the scrub, hacking away at the vines while looking for outcrops of rock hidden beneath the undergrowth. I could not wait to get the job done, graduate, and start working in the desert, with plenty of rocks and bugger all trees.

EARTH, WATER, AIR

It is often said that rainforests are the lungs of our planet[55]. I have read that they host more species of plant and animal life than all other

---

[55] Although in truth that title probably belongs to all the algae (phytoplankton) in the oceans.

environments on earth put together. After spending so much time in tropical rainforests in Papua New Guinea and, especially, Indonesia, I can well believe it. Life is at its most irrepressible in a rainforest. But that does not mean life is easy. Competition is intense, versatility is a pre-requisite and vigour is the key to success.

A tree falls in the rainforest, creating a gap in the canopy that allows sunlight to reach into the understory where deep shade normally prevails. Suddenly there is a burst of new growth. Plants once docile and anaemic erupt into unrestrained exuberance in the race to colonise the open space. Around the edges of the clearing curly tendrils of vines reach out, seeking a solid base from which to hang. Scarcely noticeable beforehand in the dim confines of the forest floor, creepers and soft fleshy shrubs stand up and shout "LOOK AT ME", enjoying their day in the sun, behaving as though it will go on forever. But it doesn't. For soon those lush green shoots are pushed aside by new and sturdier rivals that reach up, desperate for the sun, behaving as though it will not be there forever. A winner emerges triumphant; the place-getters gather round; the gap closes; the light fades; the wound heals and the rainforest canopy is once again whole.

Walking in a rainforest is a strongly sensorial experience. Sight, of course, has the most immediate impact, with so much variety in form, size and habitat. Plants that would startle in isolation blend into a verdant tableau as they compete for attention. Size matters in a rainforest. Height means light, somewhere up there above the canopy, so trees grow tall, anchored by giant roots that spread out across the ground with scant regard for their neighbours. Other, more insecure individuals are buttressed by aerial roots that drop down from their branches and enter the ground, bonding heaven and earth strongly, as if each were afraid of falling. Vines as thick as fence posts coil around high branches and hang lazily down into the undergrowth, offering a hand up to creepers as thin as fencing wire that emerge from the bush and climb upwards, ever upwards, seeking light and nurture. Ferns, stag horns and philodendrons abound, some with leaves a metre across that rob the light from smaller, less expansive folk. I could almost hear a sigh of relief whenever I cut one of those giant leaves to use as an umbrella, giving the lesser plants below an unexpected chance for stardom.

Rainforests are also very much olfactory environments. Strongest of all is the smell of impermanence, expressed as the damp sweet aroma of a rotting log or the burnt, organic odour of festering leaf litter, like pine nuts left too long under the griller. But slipping through the cracks of pervasive decay are softer, more subtle fragrances, some sweet and alluring, some rancid and repulsive. Flowering plants are strongly scented, for colour alone is not enough to attract pollinators in the gloom of the understory. Rocks covered in soggy moss exude an odour that reminds me of a colleague's halitosis. Colourful fungi clinging to tree trunks or jutting out in layered arrays from fallen logs smell dry and musty. Here is a pitcher plant, its giant cups filled with sticky fluid that is peppered with unfortunate insects attracted by its sweetly seductive scent, their corpses waiting to be devoured by the hungry host. There hangs a vine, its wrinkled bark smelling sour and looking as dead as those insects. But when cut the vine dribbles sweet water that satisfies the thirst of a passing geologist.

Sound is ethereal in a rainforest, for no matter how great its intensity, the source is rarely observed. From the roar of an approaching downpour to the soft tinkle of a brook hidden beneath the undergrowth or the drip, drip, drip of saturated leaves, water is a constant companion that chants a rhythmic vesper as it lubricates the forest. Animals are heard but not often seen. Here a snorting pig scratching in the undergrowth, there a monkey, chattering away to his fellows, or warning them of my approach. Down by the creek a croaking frog can be heard but is hard to find, such is the excellence of its camouflage. After rain, the atmosphere is thick and sticky, feeling as if it's never been cold. The saturated air is resistant to motion, like wading through surging surf. High in the branches the cicadas sing with raucous harmony in celebration of the returning sunlight, hot and burning. The canopy is reenergised as I ponder, how can such small creatures make so much noise?

In Indonesia the most startling sound in the forest is the whoosh of a hornbill's wings, as it flies above the canopy, unseen and high over my head. It is a primordial, unnerving sound, more in keeping with the legend of Sinbad and the Roc than a daily encounter in the forests of Kalimantan and Sulawesi.

Colour is expressed subtly in a rainforest, except of course for green, the colour of life, which assails the eye in a thousand shades and becomes a kind of botanical white noise. Less striking than the greenery is brown, the colour of death, found mostly on the forest floor, where expired leaves lie rotting as they return their nutriments to the earth beneath. Even in death the leaves give life, as beetles, millipedes and other tiny creatures crawl through the debris and devour the feast it provides. Other insects flourish in the forest but seem happy to leave me alone. Butterflies flit through the understory, looking for a mate not for me, while shiny black beetles sound like buzz-bombs as they fly purposefully across my path, as though on a military mission. Mosquitoes, fortunately, are rare in the forests of Indonesia and Papua New Guinea, preferring the watery habitat of careless humans in and around towns and villages. Not so leeches; waiting patiently on leaves moistened by recent rain, crouching like an omega (Ω), they are ready to reach out and attach silently to a passing mammal; all too often that mammal is me.

Flowering plants are common enough but, despite their bright colours, can be almost invisible amongst so much visual clamour. Orchids were always of special interest to me, as a grower of these plants at home. Most are epiphytes, clinging to trees just below the canopy, seeking light and moisture. It is only when a tree falls that orchids are easily seen at ground level, where their tenacity and vivid spectral hues can be admired by passing human eyes. Some, especially the vanda species, display bold reds, pinks and purples, colours that attract attention up near the canopy but seem extravagant in the subdued light at ground level. Do these forcibly relocated orchids survive down here in the semi-darkness, away from the strong but filtered light that nourished them above? I suspect they do – orchids are among the most resilient of all plants in the botanical kingdom.

So it was that, far from avoiding rainforests, I spent much of my geological career traversing these crucibles of extravagant life. Born of the EARTH, nurtured by WATER, the forest returns favour by exhaling clean, fresh oxygen, newly recycled into AIR for all creatures on the planet. My job was to study the rocks that lie beneath the vegetation and sustain it. It was heavy going, physically challenging work, extracting a toll of sweat and blood, but it was not without reward. For as I laboured beneath the canopy I could not help but absorb the energy that emanated so forcefully from the living, breathing forest that surrounded and embraced me. Humans are an addendum in that environment but are not alien. We are visitors not intruders. Unnecessary but not unwelcome. While not a critical part of the ecosystem, we are able to participate in it in a deeply satisfying way. In doing so we become part of the ebb and flow of life at its most exuberant, life that surrounds us, watches us but is of necessity indifferent to our fate in the struggle for survival. But we cannot be indifferent. The need to preserve and protect this critical environment is both self-evident and self-rewarding. For it is by subjugating ourselves to the aura that permeates the rainforest that we become part of this planet's most profound expression of the irrepressible force of life.

33

MOUNTAINS OF FIRE

FIRE

Volcanoes are a source of enduring fascination for me and they have frequently been the objective of my travel, both professional and leisure. I have visited many volcanoes, including some in New Zealand, Iceland, Russia, Japan, Italy and the Greek Islands, as well as North and South America, Hawaii, Papua New Guinea and Indonesia. All have left an indelible impression on me and served to reinforce my view of the earth as a dynamic, self-renewing system that operates on a timeframe beyond our imagination. Until recently, humans have had no part in this continuing drama, being little more than the flotsam and jetsam of earth's evolution. But with the emergence of modern technology and the development of nuclear weapons, the human race has become an active player in the evolution of this planet. For better or worse, and hopefully for better, we can influence outcomes and perhaps change the course of earth history. But we cannot prevent earthquakes, or even predict them very well. And we cannot stop volcanoes from erupting wherever and whenever they will. Just as well, for our very existence is dependent upon the continued activity of the earth's most visible expression of the vibrant forces beneath our feet – the volcanic eruption.

Most people see volcanoes as remote spectacles, eliciting awe and wonder when they erupt but detached from their daily life, a threat only to those who live near them. Of course, human nature has a short-term memory problem, soon forgetting the volcanic events of the past. Just about everyone knows about the destruction of Pompeii by Vesuvius in 79 C.E. but that has not stopped Naples from spreading up the slopes of 21st century Vesuvius. When the events of 79 C.E. are repeated, as they surely will be, Pompeii will pale into insignificance compared with the devastation to be wrought upon the vast sprawl that is the city of Naples today.

New Zealand's largest city, Auckland, is built upon a group of volcanoes that have erupted, on average, about every 5,000 years for countless millennia. The last eruption took place around 600 years ago (Rangitoto Island in the harbour) so, statistically, the Kiwis are safe for a while yet. But averages are just that and the demise of Auckland could be in the news next week, next year, or next millennium. A display in the Volcano Room of the Auckland Museum makes that point very vividly, with an alarming representation of the end of Auckland, complete with mocked up television news coverage showing CGI images of an eruption in the harbour and simulated earthquakes underfoot. A sign outside the display warns that:

"This display may disturb some people."

It certainly disturbed me!

Australians should not feel too smug either. Visitors to Mt Gambier in South Australia who stand and admire the city's beautiful Blue Lake can probably appreciate its volcanic origins but feel safe because it was all so long in the past. Well maybe not! Mt Gambier was formed by an eruption about 5,000 years ago and if Auckland is any guide, a renewed outburst may be imminent in the southeast of South Australia.

It is true that volcanoes are a threat to humans in many parts of the world but they are also a boon, for we are quick to take advantage of the fertile soils that volcanic rocks, both new and old, bestow upon the land. This is highly beneficial to humanity, although such fertility is hardly essential to our very existence. But there is more to the story. Volcanoes truly are vital to our ability to inhabit this planet for a very simple but fundamental reason – volcanoes are the ultimate recyclers.

There is a great deal of attention these days given to the presence and concentration of carbon dioxide (CO_2) in the earth's atmosphere. Rightly so, because CO_2 contributes to the "greenhouse effect" of the atmosphere, which is what prevents the earth from becoming an ice ball. Too much CO_2, however, would be a concern because the earth may become too warm for our liking. That, of course, is the essence of the so-called global warming debate. There is no doubt that human activity in modern times is adding to the accumulation of CO_2 in the atmosphere, where it may increase the greenhouse effect beyond what is comfortable for us. Combustion of fossil fuels is clearly the major source of anthropogenic CO_2 but population growth and the clearing of forests are also significant factors.

Sixty years ago, there were 2.8 billion people breathing on earth; today there are over 7 billion breathers, plus all their sheep, cattle and other domestic animals, to say nothing of their machines. Humans and their animals, like forests, are repositories of carbon and are key parts of the carbon cycle. Each of us exhales about 44 mg of CO_2 every time we breathe out. On average, humans take a little over 8,400,000 breaths per year. That adds up to something like 2.3 billion tonnes of CO_2 added to the atmosphere annually by human respiration. And of course, several billion tonnes more is added by our farm animals. But before we chastise ourselves for having the temerity to breathe, we should remember that the carbon dioxide we exhale originally entered our bodies as food. That food consists either of plants, that grew by extracting CO_2 from the atmosphere directly, or animals that eat plants and thus sequester CO_2 that we then absorb indirectly. It is human activity, especially the burning of fossil fuels, rather than human life, that is the major source of anthropogenic CO_2. Even so, it is hard to quantify just how much of an impact human-generated CO_2 has.

The main reason for this uncertainty is that there are so many natural sources of the gas, few of which are well quantified. Some scientists claim that anthropogenic CO_2 is as little as 3% of the total flux of the gas that enters the atmosphere each year. Others argue that we contribute a much higher proportion. Again, some researchers state that the contribution by CO_2 to the total greenhouse effect in the atmosphere is minor and non-linear; i.e. doubling the CO_2 concentration does not double its greenhouse impact. I confess that I don't know who is right or wrong, but this I do know: Our world is a great deal more complex and changeable than it appears to the uninformed or superficial eye.

For one thing, the principal greenhouse gas is not CO_2 but water vapour; it accounts for about 70% of the atmosphere's blanketing effect. This is easily demonstrated by considering the difference in overnight temperature for a given latitude between coastal and interior (desert) locations, which have the same CO_2 concentrations in the air. On the coast, the air is moist and effectively restricts the escape of daily warmth when the stars come out. In the desert, the air is dry and the warmth is radiated back into space relatively unimpeded, causing the familiar chill of early morning at Uluru in June. Even so, it would seem wise to limit the evolution of anthropogenic carbon dioxide as much as possible, at least until we understand its impact better. To call the CO_2 we emit "carbon pollution" though seems a bit harsh, given that this gas is essential to the maintenance of life on our planet.

Without CO_2 in the air there would be no photosynthesis and thus no plants; without plants, there would be no oxygen and thus no animals. Which brings me back to why volcanoes are so vital to our existence.

We are all familiar with the decay of organic matter as a natural process that recycles nutrition into the soil and allows new organic forms (animal and vegetable) to have their day in the sun (as well as releasing CO_2 into the air). Rocks decay too. It takes much longer for them to do so, but the prolonged exposure of rocks to the water and air that we encounter daily causes the minerals in those rocks to react chemically and change to compositions more stable in the oxidising environment of the earth's surface. This is called weathering. Chemical reactions during weathering extract carbon dioxide from the air, rain and surface water, commonly forming carbonate minerals such as calcite ($CaCO_3$) and siderite ($FeCO_3$). These minerals lock up CO_2 for the long term. In due course, erosion by wind and water breaks down the weathered rock into small fragments (sand and silt) that are washed into the sea and deposited on the sea floor as sediment. Some of this carbonate-bearing sediment reacts with the sea water, helping to keep it alkaline. Continuation of the process over geological timeframes buries the sediment and creates new rock – sedimentary rock, such as the sandstone on which Sydney is built, some of which is cemented by calcite. Other sediments are principally composed of calcite, forming the familiar rocks we know as limestone.

Erosion has been going on for as long as there has been land above sea level and CO_2-consuming weathering for at least the last 2.4 billion years. Why, therefore, is there any carbon dioxide at all left in the atmosphere? The answer is simple: Volcanoes recycle carbon dioxide. Large amounts of CO_2 issue daily from active volcanoes and geothermal fields around the world, including many that are unseen beneath the oceans, where it is said 85% of the world's active volcanoes reside. Along the Kermadec Trench[56], for example, liquid carbon dioxide has been observed pouring in vast amounts from vents on the sea floor – the CO_2 is liquid because of the great pressures in water depths of as much as 10,000 metres.

This deep ocean trench has been formed by the subduction of the Pacific Plate sea floor (with its carbonate-bearing sediments) beneath the Indo-Australian Plate. At depths of around 100-200km, which is below the crust and well into the mantle, the sedimentary rocks of the down-rafted plate begin to melt, generating magmas containing dissolved CO_2 and other gases. Those magmas periodically rise and reach the surface to erupt as lava and ash or crystallize in the crust below the surface. In the first case, CO_2 that was locked up in the sediments is liberated directly as gas evolving from the exposed magma. In the second case, the gas escapes from the trapped magma into the overlying rocks and permeates to the surface, where we find it venting in geothermal fields like Rotorua in New Zealand. That process has been adding to atmospheric CO_2 since the dawn of geological time. It has also been replenishing the CO_2 extracted by weathering at least since the atmosphere first contained oxygen, around 2.4 billion years ago.

Thus, we can say that for as much as subduction zones, and the volcanoes, earthquakes and tsunamis they give rise to, are to be feared for their destructive potential, they are also a prime reason we are here at all. The harmony and balance of nature is not restricted to the animal and vegetable kingdoms. Armed with this knowledge, visiting volcanoes has become something of a pilgrimage for me, a passion that has taken me to some remote corners of the world.

---

[56] A linear deep ocean trench that runs from New Zealand to Tonga.

34

MUTNOVSKY

FIRE

"I can't say it's stable ... especially in wet weather it can be very slippery, quite dangerous I can say and there are many rocks you can see at the head. But this is the place. I don't want to think about danger ... I read many books ... they say you should think about something good."

Leana, our guide, spoke earnestly as we stood inside one of the several craters of Mutnovsky volcano. I shared her enthusiasm and felt very keenly the reverence for place that emerged from her engaging, Russian-accented English.

Danger and exhilaration are the odd couple in many of life's landmark experiences. As I listened to Leana and looked around me any sense of impending peril was swamped by the thrill of being there. It had been a few years since the last eruption and as sure as more will come, there was no sign that a new fiery outburst was imminent.

The Kamchatka Peninsula is the easternmost part of Russia, just across the northern Pacific from Alaska. It is part of the Pacific "Ring of Fire" and is one of the most volcanically active regions on earth. The peninsula is about 1500km long, between 100 and 400km wide, and is joined to Siberia at its northern end. More than 300 volcanoes have been identified in Kamchatka; most are extinct or dormant, but the Russians consider that 29 of them are active, Mutnovsky being one of them.

Getting into the Mutnovsky craters requires commitment. To begin with, you have to get to Petropavlovsk-Kamchatsky, the principal city of Kamchatka. It is not exactly on the way to anywhere else. We got there via Seoul in Korea and then Khabarovsk on mainland Russia, before flying across the Sea of Okhotsk to Petropavlovsk (Russian for Peter-Paul).

A day trip to Mutnovsky starts with a 70km, 4-hour drive in a converted six-wheel-drive army truck. The first 50km are easy – proper roads, first sealed then unsealed, are swallowed up effortlessly by the truck, less so by the passengers bouncing around in the back. But then the stakes are raised – the last 20km take 2 hours and involve seemingly endless cross-country driving, crawling at snail's pace through a weird landscape of volcanic cinder cones, ash deserts and lava fields. The driver follows a vague track that shows no respect for travellers as it rolls and winds over hill and gully or between car-sized blocks of rock ("volcanic bombs") ejected by recent eruptions. Sitting sideways in the back of the truck, with no seat belts and very little to hold on to, it takes great stamina just to avoid injury.

Some four hours after leaving the city our group of a dozen Aussie tourists arrived at the end of the trail, still several kilometres from our ultimate objective. As we emerged from the truck onto a desolate, rock-strewn plain, it was somewhat comforting to see that we were but one of six or eight such groups, all looking decidedly wobbly after four hours of bum-bashing torture. The first need was to find somewhere to pee – there were no trees to hide behind so, decorum abandoned, we took it in turns to crouch behind one of the larger rocks nearby. While this need was being met, the driver erected a table and set out our lunch; at least in this remote location we were spared the fish soup that is otherwise the *plate du jour* in Kamchatka. As we munched on muesli bars and cold chicken our eyes drifted up to a thin line etched on the slope above us.

"That is the track we must follow to get into the crater," announced Leana.

"Shit, I'll never make it up there," exclaimed the Bogan.

"Me neither," said his equally rustic wife.

Several of our other fellow travellers were also daunted by what they could see ahead of them and opted to stay where they were while we intrepid ones, seven in all, began the hike up and into the most accessible crater, which was out of sight, over the skyline. For the first fifteen minutes we traversed the rocky plain, rising slowly, until we came to a large snow bank left over from the winter. Traversing the snow was fairly easy going, as our feet followed the path compacted by previous visitors. After the snow, the ground sloped steeply upwards and the track switched back and forth to make the ascent to the skyline ridge more feasible. It took a lot of effort to get there, but once gained, the expansive view from the top seemed even more bizarre than the terrain we had been driving through earlier. Volcanic cones, deep gorges, ice fields and rocky, treeless plains stretched away into the distance. The view could hardly be called beautiful in a conventional sense, but laid out before us was a wild, brutal landscape of nature stripped bare.

On we trudged, following Leana, down a crumbly talus slope to the river that gushed out through a breach in the crater wall. Fortunately for us hikers, the river was bridged by ice and the next half hour was spent crunching along on the ice beside the torrent that was fed by melt water from within the crater. The summer thaw had crenulated the surface of the ice, which was speckled with fine rock fragments that had concentrated in place as the ice melted. These, I could see, were volcanic ash particles from recent local eruptions that must have blanketed the area at the time. Leaving the ice, we ascended again along the narrow track that wound between large rock boulders and other lithic debris, some of which had certainly rolled down from above – a significant hazard for those bold enough to take this trek.

And then we were there. What a dramatic scene faced us! In the foreground, numerous vents ejected steam and boiling water. Bright yellow sulphur was crusted around the openings, hinting at the sulphurous Hades that lay beneath our feet. Clouds of condensed steam enveloped us as the air roiled and eddied within the crater. As though exhaled by some subterranean beast with severe halitosis, each cloud of steam brought whiffs of hydrogen sulphide (H_2S), with its unmistakeable smell of rotten eggs. H_2S is highly poisonous but fortunately its odour is so intense that humans can usually detect the gas at concentrations of 5 parts per billion, 1,000 times lower than its toxic threshold of 5 parts per million.

Scanning the inside of the crater wall I could see many repetitions of the solfatara field in front of me, as plumes of condensed steam rose from countless vents and fissures in the ground. The rock itself was streaked and stained with colour: the yellow of sulphur, red and brown from oxidised iron, green where algae clung precariously to the slope and stark white where acidic waters had bleached the rock. Incongruously, sheets of ice lay between many of the active vents, lending veracity to Kamchatka's claim to be the Land of Fire and Ice.

We were all totally awe-struck by the raw power of the place. Surrounded as we were by the near vertical crater walls there was a strong sense of entrapment. We were not looking into it from a safe distance but were right inside the beast, clinging to its vital organs. I watched with some concern as Brett, our intrepid tour leader, ventured too close to some of the steaming vents for my liking. The danger was real – he could easily have broken through the crust around the opening and been severely scalded. But …

"I don't want to think about danger," Leana had said, expressing proprietorial authority as she stood with us inside the bubbling cauldron.

It was not my first experience of active volcanism, but its impact on me was as enthralling as it was intimidating for our group of volcano neophytes. I observed their wide-eyed expressions and listened as they tried to give words to this alien experience:

"Primeval."

"Infernal."

"Hell on Earth."

"Apocalyptic."

"Nonsense" was my reaction to all of that.

This was geology in action; evolution right before our eyes; renewable energy *in extremis*. We were privileged to witness there an uninhibited display by the vibrant earth that spawned us. Visiting Mutnovsky is a dynamic, multi-sensorial experience, one that I responded to at the time by thinking:

"The colours of juxtaposed rock and ice please my eyes, while the roar of escaping steam pulses rhythmically in my ears and balmy malodorous vapours caress my skin. Here the earth is alive! I can smell its breath; I can hear its heartbeat; I can feel its warm embrace."

We stayed for as long as we could, absorbing the scene around us and allowing our senses to fix an indelible image into our brains. The return trek was uneventful, even anti-climactic, and the drive home as darkness fell was tedious and tiring. Later that evening, as we de-briefed over a beer or two, or, in the Bogan's case, a vodka or three, there was a strong sense of camaraderie, of shared adventure, of danger averted and safety regained. Like Dante, we had been into the Inferno and led, not by Virgil but by the lovely Leana, we had survived to tell the tale. I will not pretend that my simple prose will compare with Dante's epic, but I will remember Mutnovsky and be thankful for the experience it gave me for as long as I live. Perhaps it will even prepare me for what may come after that!

The only way to get around in Kamchatka

Inside the crater at Mutnovsky in Kamchatka

Valley of the Geysers, Kamchatka

35

VALLEY OF THE GEYSERS

FIRE & WATER

The big *Mil Mi-8* helicopter lifted off from the airfield and turned northwards. On board were two Russian pilots, our Russian guide, Leana, and a dozen intrepid Aussie tourists, all of us bound for the Valley of the Geysers, a UNESCO-listed World Heritage Area, about 200km northeast of Petropavlovsk in Kamchatka. As we cruised along at low altitude we were treated to one of the most stunning visual spectacles that nature has to offer. As far as the eye could see, countless volcanoes popped above the landscape, like pointy mushrooms after rain. The smaller volcanoes were capped by bare, rocky accumulations of lava and ash. The larger mountains were snow-capped and many had glaciers hanging on their flanks. A few stood head and shoulders above the rest, most of all the 4500m, ice-capped, perfect cone of Kronotsky, jutting skyward away in the distance.

Closer to hand the ground below us was pristine sub-arctic wilderness. Wild rivers and creeks flowed through the landscape of tundra and birch forest, with no sign of human habitation. Not a road, not a track, no cultivation, not a building of any kind. Once or twice the pilot brought the helicopter down low over a large river, allowing us to see huge salmon swimming upstream on their way to spawn and die.

After 45 minutes or so we reached Karymsky, a continuously active, regenerated volcano that has grown on the margin of a large, lake-filled caldera (where a much larger volcano once stood). As the helicopter performed tight circuits around Karymsky we stared out the window at the awesome sight of the volcano fuming strongly. Billowing clouds of steam and some ash issued from the crater and blew across the adjacent lake. Down the flanks of the mountain recent lava flows coalesced into a valley that channelled the lava around the base of the volcano and towards the lake.

Next we flew over the lake-filled crater of Maly Semiachik. Inside that crater the water was a pale, bleached, creamy colour and clearly very hot, as clouds of condensed steam could be seen rising from its surface.

"That lake is so acidic it would dissolve an iron bar placed into its water," Leana explained.

Finally, after an hour and a half of flying, we reached our first landing: the UNESCO-listed Valley of the Geysers. Gently, the big chopper settled onto the helipad and we all exited through the side door, instinctively lowering our heads as we ran away from the still turning blades; not that there was any risk – the helicopter was so big the rotors were well above our heads, but it still felt better to duck. Minutes later a second identical aircraft arrived with another load of tourists and put down onto an adjacent helipad. Once both machines were shut down and the whine of turbines subsided a new sound invaded our ears – the roar of steam escaping under pressure. Over the next hour or so we walked slowly along a prepared pathway that took us past countless pale blue boiling pools, fuming steam vents rimmed with yellow sulphur, erupting geysers, trickling hot springs and burping, bubbling, boiling mud pots whose colour looked artificial, like melted caramel. All of the thermal phenomena are driven by magma at shallow depth below the caldera that contains the valley. Drilling has shown the temperature 500m below the surface to be 250°C.

"There are around ninety geysers within the valley," Leana told us, "spread out over a distance of 6km. This is the second largest concentration of geysers in the world."

The most impressive geyser we saw was located right on the banks of a lake, formed when a giant mud flow dammed up the Geysernaya River in 2007. A geyser is generated by a rising column of expanding superheated water that does not boil initially because the high pressure at depth raises the boiling point of water substantially. As the column rises further, however, the incumbent pressure drops until suddenly the boiling point is below the ambient water temperature and the superheated water flashes to steam. The steam then generates its own pressure, pushing the column of water above it upwards and out through the opening of the conduit. Once the water column has been ejected it trickles back into the conduit, filling it up and allowing the cycle to continue. Some geysers cycle every few minutes, while others take days or more. In Kamchatka, the geysers we saw had short cycles, many just three or four minutes, so there was plenty of opportunity to photograph them in action.

All too soon (at least for me) we were bundled back into the helicopter and took off for the short flight west to Uzon Caldera. Inside that caldera, the ground was flat and nearly treeless and the many hot springs flowed into ponds and small lakes that were surrounded by bleached and stained soil and rock that gave the whole area an exotic technicolour palette. Next to the helipad was a small wooden cabin used as a summer base by the local guides, one of whom had greeted us on landing. About 50m from the cabin stood a small toilet building, like a lonely porta-loo. Several of our group decided they needed to use that facility, until, as they approached, a huge brown bear[57] stood up from amongst the blueberry bushes it had been grazing on. My companions' needs suddenly vanished as they made a quiet but hasty retreat to the cabin, where the local guide stood with a Kalashnikov at the ready.

A short walk through the thermal field ensued. The most striking feature of this location were the mud volcanoes – miniature cones, just 10-20cm high – which every few seconds ejected little vomits and dribbles of light grey mud, like overindulged anaemic gluttons. Then it was back into the chopper and as we took off I spied the bear, still grazing nonchalantly and seemingly heedless of the machine rising from the ground nearby in a flurry of noise and wind.

In my travels as a geologist I have been fortunate enough to have nestled into some beautiful, even spectacular scenic spots for a lunch break. But few could match the vista presented to me that day. In clear blue skies, light winds and warm sunshine, the Mi-8 touched down on a bank overlooking the Karymsky caldera so we could enjoy our cut lunches in the fresh air. We were hungry and it was already about 2.00 pm, but it was hard to concentrate on eating as we looked out across the bright blue lake at the perfect cone of Karymsky volcano, now fuming more quietly than it had been that morning.

"In 1996, the lake was being used as a fish farm," Leana told us, "when a big eruption occurred and lava was erupted under the lake, causing all the water in it to boil."

"It must have been the biggest bowl of fish soup in the world," I quipped.

At least the Bogan thought it was funny!

***

---

[57] Related to the Kodiak or Grizzly Bear in Alaska

Our final stop that day was Nalychevo Natural Park, where we left the helicopter and moved into a couple of dormitory-type basic cabins (one for the boys and one for the girls) for a two-night stay. It was a true wilderness camp, with only the most basic of facilities, a separate hut for cooking and eating (cook and food supplied) and pit toilets, but no showers. This was thermal country too and our ablution needs were met by three seriously hot pools located some 400m from the cabins, right next to a small rivulet. In the hottest of the pools the water temperature was probably at least 45°C, with the other two slightly cooler and more tolerable. In contrast, the river water was near freezing. What a shame the river was too shallow to dive into after a hot immersion! Trekking across to the pools was quite nerve-wracking though, as bears had been sighted nearby a few days before. Fortunately, the tundra was quite shrubby and low so we could see for a fair distance over the top of it. On our bathing excursions we kept a wary eye open for bears and I took comfort in being, if not the fastest runner in the group at least faster than the slowest, so survival seemed likely.

As we floated in the hot pools, using the algae in them as exfoliates for our skin, I reflected on the exoticness of our location. A clear pristine river sparkled in the sun next to us. Further away the distant views were framed by mountains, including many barely eroded, cone-shaped volcanoes, most of them capped in snow and ice. The tallest of them, Koryaksky at 3456m, was not only white-capped but displayed two long, narrow glaciers hanging down the north face towards us, joined at the top like a pair of frigid trousers. A steady jet of steam rose into the air from a vent high on the western side, testament to the thermal energy hidden beneath the mountain. Luxuriating in the warmth and comfort of the hot pools it was hard to believe that in six months' time up to 15m of snow would lie on the ground at Nalychevo.

After a somewhat uncomfortable night on a bunk in the cabin I arose to a sunny, bright morning, with no wind to rustle the birch trees. Koryaksky to the south was etched against the blue sky, its ice cap glistening in the morning sun and the steam vent bigger than ever in the brisk morning air. While half the group stayed behind in camp for the day, the rest of us set off soon after breakfast for a 25km hike through the wilderness. Leana was once again our guide but we were comforted by also having Sergei, from the Nalychevo staff, come along with us. Not only was he an extra guide but he also carried a Kalashnikov that he brought along to protect us from any rampaging bears we might encounter. Fortunately, or perhaps unfortunately, we did not see any bears.

What we did see were countless vistas of volcanic mountains, dripping birch forest and pristine tundra, interspersed with fields of blueberry bushes, laden with ripe fruit, as appealing to us as it was for the bears. Our destination was a very active group of springs, where cold CO_2-laden water bubbled from the ground in large quantities. A small hut had been built over one of them, complete with a dipper to reach down and sample the naturally carbonated water. A multi-lingual sign inside the hut said:

"Do not sleep inside this hut or you may suffocate because of the high concentration of carbon dioxide in the air."

The hike was long and tiring, especially on the return leg, but what a privilege it was to trek through such unspoiled, untamed sub-arctic wilderness. It was very gratifying to know that such places still exist in this overpopulated and widely polluted world.

The next day another Mi-8 helicopter arrived at Nalychevo and picked us up for the short flight back to Petropavlovsk.

With our Russian escort, Sergei, and his Kalashnikov at Nalychevo Park

The lava fountain at Mauna Ulu, Hawaii, in December, 1973

The same lava at Mauna Ulu in 2011

36

HAWAII

FIRE

"Folks, this is the Captain. We'll be arriving in Hilo in just over an hour. The weather in Hilo is clear and we should have you at the terminal on schedule at ten after five."

The Captain's voice awoke me from my in-flight dozing and I began to think about where I had been and what lay ahead. I had just spent twelve weeks in the United States, based in Salt Lake City and visiting copper mines and prospects owned by my employer, Kennecott Copper, in Utah, Arizona and New Mexico, as well as Kennecott's research laboratories in Massachusetts. My boss in Sydney, Dr Richard (Dick) Nielsen, who became a long-term friend and mentor, had sent me there to learn at firsthand about the geology and exploration of the large, low grade copper deposits known as porphyry copper deposits. These were the prime target of Kennecott and most of the other big copper companies, such as Anaconda, Phelps Dodge and Rio Tinto[58]. Exploration for this type of deposit had been occurring in North and South America for decades and had spread more recently to the Southwest Pacific, following the discovery of the Ok Tedi and Bougainville deposits in Papua New Guinea.

Dick had sent me to the USA in September, 1973, just two months after I had been hired to his staff in Kennecott's Sydney office. He

---

[58] BHP was not a big player in the copper business at that time.

said he wanted me to be as well-equipped as possible for the role that lay ahead of me, which included porphyry copper exploration in Indonesia.

"An extended field trip to all the classic porphyry copper areas in the Southwest of the US will be just the thing you need to get you up to speed," he suggested.

"I'm up for it," I replied, with scarcely concealed fervour.

"Well I'm not so sure that such a long trip is a good idea. I've got Julianne to look after and I'm five months pregnant with the next one." Margaret, my wife, was less than thrilled about the prospect of my long absence. At least we had just moved into the house we had bought in Gordon, on Sydney's North Shore, so she was getting settled, although there was still much unpacking to do. To her great credit, she acknowledged the need for me to go for professional reasons and did not stand in my way. This was the second time Margaret had deferred to my professional interests, the first being my ten-month sojourn in Berkeley before we were married. Nor would it be the last. Not many women would have been so generous and I felt a heavy obligation to make the most of my time away.

There is no doubt that I did learn a great deal during the site visits and by interacting with geologists already working in the porphyry copper field. Visits to operating mines gave me firsthand experience of the giant scale of these deposits and a sense of what was needed when evaluating them as raw prospects. The knowledge I gained set me up for what became a major focus for much of the remainder of my career. Salt Lake City had been an intriguing place to be based, too, with its dominance by the Mormons and isolated, desert environment, overlooked on the east by the lofty Wasatch Mountains. The city itself was laid out in a rectilinear grid. Kennecott's office address was 1700 South 2300 West in Salt Lake City, which meant that the office was 17 blocks south and 23 blocks west of the city centre. Pre-GPS navigation was dead easy. The Mormons were nothing if not well organised!

"Folks it's the Captain again. I've just been told that a new eruption of Kilauea volcano has started this afternoon. I gather it looks quite spectacular."

Again, the Captain's voice interrupted my reverie and my mind returned to the present. I had decided to make a detour on my way home from the USA to visit the Hawaii Volcanoes National Park on the Big Island, Hawaii. Accommodation had been booked at a hotel within the Park and tomorrow would be dedicated to touring in my rental car through some of the classic examples of recent volcanism that Hawaii presents so well. Now, as I had just heard, there was the prospect of seeing an actual eruption taking place. What a bonus!

It was dark by the time I reached the hotel within the Park. Away in the distance I could see a red glow that I knew to be the Kilauea eruption and I was anxious to see it up close that very night. After dumping my luggage in my hotel room and gobbling down a quick bite to eat, I returned to my rental car and headed off in the direction of the red glow. Shortly afterwards I came to a group of cars parked beside the road, which had become a dead end when covered by a lava flow sometime in the recent past. I marvelled at the sight of the white centre line simply disappearing beneath the black rock. But even more marvellous was the sight of the lava fountain that was spurting into the air just a kilometre or so away.

This, the latest expression of Hawaii's almost constant eruptive activity, was known as Mauna Ulu. It was a small crater on the flanks of the much larger Kilauea Volcano and a well-defined walking track led from the impromptu carpark towards the dramatic display lighting up the night sky. With my Minolta camera and much excitement, I set off along the track. As I got close I could see that there was a wooden platform built, obviously before the eruption, on the edge of the crater, opposite to where the lava fountain was rising. Several people were standing nearby, just below the platform and crater rim. I approached cautiously but eagerly and saw that one of the people was a Park Ranger. He was there, I assumed, to monitor the eruption and decide if it was safe to let visitors like me approach so closely. Camera at the ready, I walked up onto the platform overlooking the crater.

Immediately I was assaulted by a blast of heat the like of which I had never experienced. Within seconds I understood why the other people were standing off the platform and below the rim. I would have to join them in a moment, but not before taking in the dramatic scene in front of me. On the other side of the crater, about 200 metres away, lava was spurting perhaps 100 metres skywards, before falling back into the crater. A rhythmic shushing rose above a regular deep rumbling as the lava emerged from the ground. There were no explosions and only a hint of sulphur. Clearly, very little gas was involved in this eruption, as is true of most Hawaiian volcanoes. So smooth and steady was the lava fountain that it was as though a giant subterranean fire hose had been turned on by an antediluvian fireman. Except that it was glowing, molten rock at 1100°C, not water, which was being ejected. Then I noticed a constant splattering sound as the falling lava hit the black crust that had formed on top of the lava accumulating inside the crater. After just a few seconds I retreated from the platform, happy to have snapped a few pictures.

Down below the rim I joined the group standing around the Ranger and listened to his spiel as he explained the history of this volcano to the visitors. All the while the eruption continued, just over his shoulder and across the crater. I went back up to the platform for another look, this time captivated by the lava lake within the crater. It was covered by a black crust of chilled rock, broken by numerous cracks, like the outside of a crusty loaf of bread. The cracks glowed red, hinting at the seething cauldron of fire lurking just below the surface. Then, as if to intimidate the spectators, a section of crust turned over, revealing a glowing, ragged underside that radiated heat like a beacon, making a presence on the platform even more challenging. For the next hour or so I repeated my dashes up onto the platform, watching the drama unfold for a few seconds at a time and noting that the crater was filling up. My only regret was that I quickly ran out of film and had neglected to bring more with me.

At one point the Ranger kept us back from the rim, as a large rock became wedged in the fountain of lava, splitting it into two streams, one of which was pointed at the platform and falling uncomfortably close to us. As suddenly as it had appeared, the rock must have dropped into the lake, or perhaps melted, and the fountain resumed its vertical trajectory. Sometime after midnight I reluctantly began to walk slowly back along the track to where I had left the car, turning around every few minutes to take another look at the display. For someone who had studied volcanic rocks for his PhD, who had visited young volcanoes all over Western America, not to mention Papua New Guinea, this was a unique and very fulfilling experience. A life-long fascination with volcanoes was etched even more deeply into my soul.

The next morning, rising a little late I confess, I returned to Mauna Ulu. Since my visit during the night the crater had filled to overflowing and a steady stream of lava, glowing red in the morning sun, was pouring rapidly down the outer slope. All the while the glowing fireman's hose squirted straight up into the air, feeding the flow. The wooden platform that had allowed me to sample the erupting experience so closely the night before had disappeared, presumably burnt up as the lava lake rose beneath it. More importantly, the Park rangers were out in force, prohibiting any visitor from approaching closer than 500 metres from the erupting fountain.

As I sat on the rim of an older crater half a kilometre or so from Mauna Ulu, munching on a sandwich, I looked across at the continuing spectacle. It was already difficult to believe that, for one night only, the very night I was there, it had been possible to stand right on the crater edge of an erupting volcano. And live to tell the tale!

"Now this will be a story to tell my grandchildren!"

**

I returned to Mauna Ulu, by then long quiet, during a visit to Hawaii in 2011. While viewing the information provided at the Volcanoes National Park Visitor Center I chatted to one of the young Park Rangers, telling him of my first visit.

"In 1973 I was here on the very day that Mauna Ulu began a new eruption. For that one night, it was possible to stand right on the edge of the crater as the eruption was underway and filling the crater with a lava lake."

"How was that possible?" he asked, incredulous.

"Easy. There was a track from the road to the site and when I reached the crater a Park Ranger was there, keeping an eye on things."

"Well, there's no way in the world we'd let you do that these days!" he exclaimed.

37

OUTBACK ELEGY

One of the things I admire most about Aboriginal culture is the concept of "Country". As I understand it, the term means much more than just "my roots" or "where I come from". It includes the land underfoot and everything that lives in it or on it or over it. Country seems to give meaning to a traditional Aboriginal life, imparting a sense of identity and self-esteem. There is a continuum between self and country – not "I live in my Country" but "I am my Country and my Country is me".

I have seen the look of sheer delight that came over the face of an Aboriginal acquaintance when he took me to the Clarence River in northern NSW and we looked across the surging water to a bold, high cliff face.

"This is my Country," Trevor said proudly, "here I talk to my ancestors and they talk to me."

Trevor's voice quivered as he spoke to me of his Country and I was in no doubt about the profound meaning that river bank brought to him. It was a shrine without icons. A sanctum without walls. There we could eavesdrop on the lively tête-à-tête between the river and its rocks, listen to the contented whispering of the she-oaks and hear the happy squeaks of the sand beneath our feet. Those sounds were at the heart of Trevor's identity, binding him to that place as it gave him a sense of worth. It was truly a privilege to share in this spiritual experience. As I stood next to him I felt a little tingle of pleasure as

Trevor's Country began to embrace me too. From that moment on, I understood more than ever before why land rights are so important for our indigenous people. To separate them from Country is to split them asunder.

Those moments I shared with Trevor were profoundly moving because it was truly an experience shared, not just observed. For while not indigenous, I, too, am a son born of this land and I, too, have learned to listen to its voice, in the process finding my own Country. As I have done so respect for this "wide brown land" has been etched into my consciousness. Like my indigenous friend, visiting my Country has imparted meaning to my existence and texture to my soul.

Most people can admire a scenic view: A snowy alpine crest, a cliff overlooking the raging sea, a reflective mountain lake or a colourful desert landscape are easy to visualise and enjoy. Their roles in maintaining our sense of balance and identification with the sustaining earth beneath us are important but they present no serious challenges to our intellect. As edifying as those images may be, they are, in the end, just that – images. Pictures at an exhibition. Post cards on a rack. Screen savers. To experience the natural world internally, to feel its spirit pulsing in our veins, to commune with Country, we must tarry a while and be quiet. Silence allows our senses to engage with the rhythms that arise like chanted vespers from the land around us, be it dripping rain forest or desiccated desert. In such places, such pockets of County, the fleeting pressures of human life yield to the more enduring values of nature. My indigenous friends have taught me that while life for the individual is tenuous and brief, Country offers a fortitude and continuity which can imbue us with a sense of worth that is timeless.

We in Australia live in an ancient land whose temporal dimension is so great that its geographic transformation has slowed to an amble. Change in our land is unhurried, subtle and immensely patient, true as much of our indigenous people as it is of our arid landscape. Our indigenous culture has endured for longer than any other and it behoves us to examine why this is so. For I have observed that many of the conflicts that occur between indigenous and non-indigenous Australians seem to arise from differing perspectives of time. In the city, time is precious and we lust after it, claiming there is never enough. Patience may be a virtue but it is one distributed very thinly across our urban society. In the desert time is plentiful and patience is an abundant commodity, though not one without value.

It takes a commitment of time for the uniquely Australian character of our landscape, that vast expanse we call the Outback, to register in our awareness. Not all of the Outback is truly noteworthy, to be honest, for much of it is bleak, barren and a genuine challenge to survival. But as befits our national identity, hidden within our wide panorama are pockets of serene charm that require time to notice and appreciate. We need time to hear and really see the rocks and soil, the grass and trees, the insects and animals that otherwise form just a background tapestry. Country is more than beauty, and less. For it is not what we see in Country that matters but what we feel. The radiance that emanates from Country comes to us not with demands for submission but with invitations to communion. It takes time and patience to immerse ourselves deep enough into the ordinary that our senses are awakened to the extraordinary, those jewels of the Outback that link us to our land like worm-holes in hyperspace.

You can come upon them unexpectedly, in settings where, to the uninformed eye, there is no splendour, only heat, harsh light, dust and flies. They may lack the grandeur of an alpine peak or the reflective serenity of a mountain lake, but their impact can be just as great, precisely because they are so unexpected. The Mount Isa region in northwest Queensland is replete with such surprises. For much of the year the temperature hovers in the thirties, or forties, occasionally even more. The red rocks, that sporadically show splashes of green where the copper sulphides they contain have oxidised to malachite ($Cu_2CO_3(OH)_2$), stand up in barriers, dividing the land into domains of ridge and valley. Bold outcrops soak up the sun and radiate it back onto passers-by with a vengeance. Between the ridges, in the exposed valley floors, the hot wind whistles in short gusts, summoning spinning dust devils that dance across the bare soil patches, whipping up the dust but leaving the gibbers undisturbed, before disappearing into the scrub, as though embarrassed by their exuberance. The bush itself seems lifeless, leaves hanging down despondently, despairing, without hope. In reality, those leaves are just conserving moisture in the relentless solar blaze. Colour and contrast fade in the intense sunlight and distance perception is lost. It is as though the scene is viewed with one eye closed. But then, down below the ridge, secreted away in gullies that cut across the grain of the land and offer shade and sometimes even a refreshing pool, pockets of Country lie waiting to be discovered; quiet glades that provide sanctuary and total contrast from the severity of the exposed ridges and baking valleys.

I remember one such experience shared with Russell Meares, a geologist colleague, while we were working north of Cloncurry:

> *Wearily we complete the cross-country traverse, dump the soil samples we have collected into the back of the Landcruiser and climb into the front seats. Russell starts the engine and a blast of hot air assaults us as the air conditioning comes to life.*
>
> *"Look at that!" I point to the vehicle's outside temperature indicator.*
>
> *"55°! Now that's hot!"*
>
> *"It's only because the vehicle's been standing in the sun,"*

suggests Russell.

He is right. After five minutes driving the outside temperature drops back to a mere 48° C.

"I know a good spot for lunch," Russell tells me, "we'll be there in ten minutes."

We enter a narrow glade, nestled in the shadow of an ironstone ridge where a dry creek cuts through the hard rock. A ghost gum offers us a low-slung branch to sit on while we eat our lunch, its smooth bark a cool compress, caressing our bodies while it draws away the heat. In the stillness, a stumpy-tailed lizard emerges from hiding and waddles past, looking for lunch itself. The grass here is longer, lusher and more vigorous, crushed into cradles where kangaroos have rested. Zebra finches, hardly noticeable out in the brilliance of the sunlit scrub, tweet their monotonal call, enjoying the cool and shade as much as we are. The sandy creek bed shows signs of recent flow, its multi-coloured gravel smoothly rounded and stacked neatly by the energy of the now vanished water. Some stones are streaked with quartz veins – aha! Now that's interesting! My geologist's eye sees not stones but rocks, each with a name and a story, samples of all that lies upstream presented for our perusal. I drink from my water bottle as my eyes scan the ground. The ants, seemingly oblivious to the heat out there are busy here too, tracking back and forth with the determination that exemplifies their kind. Caterpillars, linked end to end in a chain of common purpose trail past my feet. I resist the temptation to swing the lead around to join the tail and form a furry Catherine wheel. A half hour of rest and this little piece of Country has done its job. We rise refreshed and ready to face the solar incandescence once again.

In the Kimberly, that part of Western Australia closer to Asia than to its southern capital, recuperative sanctums are common too. For me, many have been dry, sandy creek beds, meandering out across the plains from adjacent hills. Giant ghost gums that shun the open spaces are happy here to dip their roots into the sand and soak up hidden moisture. Residual pools of water lie on the surface, shrinking as the dry season takes hold but testimony to the presence of more water at depth and explaining why these big trees are to be found only in the creek beds. Places that seem ordinary in the glaring light of midday become restful and refreshing as the sun sinks and the balmy breezes of sunset caress the land. In the distance, the cliffs blush bright red in the low-angled light, their shape and form accentuated by shadows, transforming a two-dimensional collage into a three-dimensional diorama. Spinifex clumps, luminous and green at this time of day, dot the slopes below the cliffs, as though inspired by, or perhaps inspiring, the dot painters of the Central Desert.

There we would sit, watching this transition, boots off, toes in the sand, drinking a stubbie of cold beer and speaking in hushed tones out of reverence for the sanctity of our environment.

Many were my experiences thus; some I remember, most I forget, recalling only the feeling, the sense of belonging, the communion with Country that lingers in me still. But there was one occasion that stands apart in my memory, a defining moment in a life full of rare and enriching experiences. It happened in the Yilgarn, a wide, empty region of Western Australia stretching for a thousand kilometres northeast of Perth. To geologists, the Yilgarn Block holds a special place in the catalogue of Australian geology, for it is here that we encounter the ancient core of this vast land. Its oldest rocks were formed over three billion years ago, when the earth was young and life was just beginning. It was one of these ancient strata that drew me to the Yilgarn and enabled me to engage in such an unforgettable experience.

EARTH

The news came over the car radio as I guided our Landcruiser along the narrow track, my colleague Paul Wilson beside me:

> *"United States aircraft have bombed Tripoli and Benghazi in Libya. The attack was ordered by President Reagan in retaliation for the bombing of a discotheque in West Berlin ten days ago in which three people, including two American servicemen, were killed and many more Americans were injured. The United Nations has urged calm amidst fears that the conflict could escalate into all out war."*

"Jeez Paul, World War III is breaking out on the other side of the world and here we are in the middle of nowhere chasing a bunch of bloody emus."

I increased speed a little, drawing closer to the emus; there must have been at least a dozen of them. The emus quickened their pace as our vehicle drew nearer.

"Yeah, it does kind of emphasise just how remote and away from it all we are here."

I slowed the vehicle right down; the emus stopped running and continued at walking pace, resolutely sticking to the track beside the fence line that we were driving along.

It was the 15th of April, 1986 and Paul and I were conducting a geological reconnaissance through a remote part of Western Australia, around 500km north northeast of Perth, within what is known to geologists as the Yilgarn Block. Our objective was to examine exposures of a layered ultramafic rock unit called the Windimurra Complex. Once deeply buried but now revealed at the surface, this rock was intruded into the primordial crust, where it cooled and separated into layers, one of which, we supposed, might contain platinum. Our model was the Bushveld Complex in South Africa, of similar age and geological setting and home to the Merensky Reef, source of most of the world's platinum.

Paul was a senior member of my geological team, highly respected for his scientific skills and equally cherished for his impish humour that radiated from beneath a thick crop of curly black hair. As we understood it, the track we were following would lead us to some excellent outcrops of the Windimurra Complex where we could assess its potential as an exploration target. The emus seemed to have other ideas.

"Come on Garry, get us past these bloody birds, will you?" I looked at Paul, a huge grin confirming his amusement at our predicament.

I speeded up; the emus began to run. I slowed down and dropped back a bit; the emus stopped running and reverted to a leisurely walk, still refusing to leave the track. I tried again, same result. On our right was a barbed wire fence; on our left dense, prickly scrub. Faced with those options, I would have been reluctant to leave the track too.

"This is bloody hopeless," Paul said, as amusement turned to frustration after we followed the birds for kilometre after kilometre. "We'll never get there."

"Yeah, I've had enough of this too. Hang on."

I drove right up to the emus that were by then running at full pace. I pushed on, causing panic in the flock but forcing a resolution. In a flurry of feathers and dust, the flock broke up, some scattering into the scrub on our left, others diving through the barbed wire fence. When I looked in the rear vision mirror I could see a bird momentarily entangled in the wire before breaking through to the other side and disappearing into the bush.

"Well, that certainly resolved the issue," exclaimed Paul.

"Yeah, hopefully not too much more than pride and a few feathers were lost in the process."

We drove on, found the outcrops and became geologists again, rather than emu terrorisers.

***

In mid-April the worst of the summer heat was over but it was still damned hot. And dry. As we traipsed around the outcrops the bulldust rose in little puffs, like splashing water. The sky was cloudless, bright blue, the sun relentless and the breeze warm, sucking moisture from our skin. The black rock outcrops soaked up that sun and threw it back at us, as though in competition with Sol, while we sat, each on a separate boulder and under skeletal shade to discuss the significance of what we were seeing. Eventually, satisfied with what we had achieved, it was time to move on. We planned to camp for the night at a water bore marked on our map, ten or fifteen kilometres away. With the sun beginning to sink into the west, we headed east, diverting from the fence line track onto another that would lead us to our intended camp site.

Half an hour or so later we drove up beside the water bore, near but not too close to the old windmill slowly clanking in the warm breeze. Overhead the sky was still blue, but in the west a vivid red-orange glow rose steadily above the horizon, like an approaching swarm. To the east, a cloak of darker purple hues nibbled at the trailing edge of blue. Together, the colours of east and west swallowed the remnant radiance of another Yilgarn day. Slowly, we two weary geologists stepped out of the vehicle and stretched.

"This looks good Paul. Best campsite we've seen this trip."

Paul agreed as we looked around us at the Salmon gums and acacias that made this place almost a wooded copse. As the windmill rotated a steady stream of water flowed into the large steel tank beside it. Next to that stood a water trough, filled to the brim and the focal point of the many cattle and sheep tracks that converged on it. It was twilight and with dusk came not only relief from the heat but a revival of our desiccated senses. Gentle aromas, too subtle to survive the incandescence of daytime, began to fill the air, refreshing us. The soft, dry fragrance of saltbush competed with the earthy odour of cattle dung, while the quintessentially Australian bouquet of eucalyptus drifted in from the Salmon gums.

In this environment darkness does not so much fall as precipitate from the air, like dew. One minute the smooth bark of the Salmon gums is glowing bright red in the retreating sunlight and the next the same trees are but grey, gnarled shapes in the thickening dusk. Before the daylight left us completely we set about making up our simple camp. We would sleep under the stars, cook our food over an open fire and wash in water from the bore. Our only concessions to comfort and modernity were two camp chairs and the Engel fridge in the Landcruiser. We did have a gas lamp but used it only sparingly as its harsh white light seemed out of character with the soft tones of a desert sunset.

It was to be our last night in the bush for this trip and by then we were well practiced in the art of setting up camp and making the transition from daytime toil to night time tranquillity; all was soon in order. By the time we had finished that task the breeze had died away and the windmill stopped turning, leaving us in an utter silence that underscored our isolation and solitude. Silence so deep my ears rang, as though resonating with the air molecules hitting my ear drums. But then, as we settled into the camp chairs and gladly popped the first cans of Emu Bitter a large flock of galahs flew in to drink at the water trough. After a few minutes of noisy squabbling and much wing flapping and chirping, they took off again as one and disappeared into the evening sky to roost who knows where? Paul and I watched them go as we popped our second round of Emu Bitter.

"You know Garry, those emus were a damned nuisance ... but they make a bloody good beer!"

"Ha ha! Yeah! A city bloke'll never know just how good a beer can taste until he comes out here for a stint in the bush."

As the light faded further, more visitors arrived, in the form of a mob of big Western Greys, some of them with the heads, legs and tails of Joeys protruding awkwardly from their pouches. They, too, drank at the water trough and, keeping a wary eye on us, began to graze on the grass that grew sparsely amongst the gibbers on the ground around the windmill.

Then it was really dark, a black, silent, opaque darkness that swallowed the light from the fire, as if it too had a raging thirst that could only be satisfied by consuming radiance. There was no moon that night, allowing us to lean back and observe the stars with a clarity and brilliance unimaginable to city folk.

"What do you think that satellite up there is?" asked Paul, as we watched a remarkably bright point of light move across the Milky Way.

"Can't be Skylab," I replied, "it crashed to earth out here somewhere a few years ago. Must be another big satellite."

As we continued to watch the sky meteors occasionally slashed the galactic canvas above us.

"I wonder if any of them will hit the ground," I pondered.

"Out here they could do that and no one would know," Paul responded.

It was Paul's turn to cook and I watched eagerly as he dumped some canned vegetables into a billy of water and dropped two great slabs of rump onto a grill. Epicures we were not! Soon the smell of sizzling steak became even more intoxicating to the two hungry geologists than the alcohol we were consuming apace.

"But first we have to have our appetiser," announced Paul, as he opened a can of smoked oysters and magically produced a packet of crackers to go with them.

"Jeez Paul, where've you been hiding those?"

After dinner, with all our beer gone, I performed some magic of my own by producing a half bottle of Scotch I had been keeping for just such an emergency, much to Paul's delight:

"Ah, wonderful; there's nothing like whisky to restore a man's soul."

And there is nothing like alcohol to turn two normally practical and grounded geoscientists into deeply contemplative philosophers, musing on all that was wrong with the world and how it could be put right. We knew for a fact that World War III had started that very day and that the end of civilisation as we knew it was nigh. But we found it difficult to care. How could we, out there, in total isolation? Hemmed in by emptiness, we soaked up the serenity that arose from the ancient soil beneath our feet and drifted past in the night air. We were as detached from the worries of the world as it is possible to be. And not a reefer in sight! As the night wore on, and the contents of the Scotch bottle slowly vanished, our ruminations became more and more abstruse. Eventually, all conversation ceased as we surrendered to the silence. Slowly, as our conscious thoughts subsided, our senses began to connect directly, one by one, with the Elements around us.

In our minds we solved many of the world's problems that night but unfortunately we made no notes, so it was all lost in the bright light of day that followed. But what did survive in my memory then, and still does today, is this: How fortunate we were to be able to share that moment of deep companionship in such a remote and uninhabited place. It is this rare and profoundly emotional sense of isolation and total solitude that to my mind is the defining characteristic of this vast and empty land we call the Outback.

Vast, empty, but sometimes cruel.

The following year Paul left our company for a more senior, and better paid, position in a Perth-based mining company that had an operating gold mine at Leonora, some 600km northeast of Perth. I was really sorry to see him go, as I greatly respected his skills as an earth scientist and had very much enjoyed his company in the field and in the office. We had similar interests, similar values and similar views of the world and we laughed at the same things. Paul had become a good mate as well as an extremely competent colleague. Fortunately, I was still making regular visits to Perth and was able to spend some delightful evenings with Paul and his charming wife, Trish, swapping stories over numerous glasses of full bodied Margaret River red.

In December, 1988, Paul's company sent him and some other executives from Perth up to the mine for a Christmas party with local staff. There were some spare seats on the charter plane so several of the Perth office staff were invited to go along. The company was doing well as the mine was going 'gangbusters' in the 1980s gold boom, so everyone was in a celebratory mood. The next morning the Perth crew boarded the *Mitsubishi MU-2B-60 Marquise* at Leonora airport somewhat bleary-eyed and looking forward to the chance to catch up on some sleep during the flight back to Perth. The aircraft took off and headed west, towards the ominous-looking thunderheads of an approaching summer storm.

The MU-2 was popular for chartering as it was fast and relatively cheap to operate. But it had a fatal flaw: It was prone to icing up in bad weather. Just after 10.00 am on 16 December, 1988, VH-BBA flew into the storm cloud. At around 10.15 am the iced-up aeroplane hit the ground with great force on Sturt Meadows Station, 55km WNW of Leonora airfield. The pilot and all nine passengers were killed, Paul Wilson among them.

A few years after that, Paul's widow, Trish, took their two boys, Andrew and Michael, to the crash site. Most of the wreckage had been removed but while there they found Paul's calculator. It was a poignant but fitting reminder of the hardworking husband and Dad they had lost.

Vale Paul Wilson. I still miss you.

With Paul Wilson at a Hotel in Germany, ca. 1987

WINDIMURRA DREAMING

Colour! The sky overhead still blue
But in the west now reddening.
Dazzling radiance gives way to softer hue,
While in the east, a cloak of purple deadening
The sky, as night precipitates like dew.

Heat! Enervating, permeating, relentless solar blaze,
Ruler of the day, parching barren fields,
Driving thermal spirals in shimmering dusty haze,
Now spent, as solar incandescence yields
To this cooler, dusky phase.

Aroma! As evening with the daily blast dispenses
And each breath becomes less painful,
Perfumes sweet and subtle fill my senses.
Soft nocturnal scents, of rancid day disdainful,
Are focussed by cool ethereal lenses.

Tranquillity! But not for long,
As squawking flock alights to drink.
A flapping, chirpy, excited throng
Of feathers grey and brightly pink,
Farewells the day with happy song.

Stillness! Yet graceful movement now is found
As Western Grey comes round the bend,
Grazing sparsely on the stony ground.
Slim pickings here my friend
But you're welcome to browse around.

Stars! Popping through the void in ones and twos at first,
Sparkling in the night time sky like sequins on finest silk.
Eyes agog, I slake my thirst
By drinking in the wondrous splash of skimmed galactic milk,
Rent here and there by meteoric burst.

Solitude! Just the two of us, no other soul is hailable.
Isolated and remote but not the least bit frightening,
Even with no phone or fax, no media available.
Joyously I close my eyes and feel my spirit lightening
As wafting up from ancient soil comes peace unassailable.

Country! Whispering voice, devoid of all that's wrong.
No bricks and mortar, no need for Ivory Tower.
A place to listen to the sounds that make me strong,
That give my life its meaning and generate my power,
For here I am at home and know that I belong.

In tribute to the late Paul Wilson, friend, colleague and fellow traveller through the Elements.

38

JOURNEY'S END

"CIVILISATION EXISTS BY GEOLOGICAL CONSENT, SUBJECT TO FURTHER NOTICE"

Will Durant[59],

American historian and philosopher, 1885-1981

AETHER

On the Indonesian island of Sumatra there is a large lake, situated towards the north-western end of the island. Lake Toba, as it is known, is about 100km long, 30km wide and up to 500m deep. It is a beautiful, peaceful, even serene place, its calm waters surrounded by quaint Indonesian villages and the verdant green of the Sumatran rain forest. Seventy four thousand years ago, the scene was rather different, for it is here that the largest volcanic eruption in human history took place.

Some well documented modern eruptions, such as Mt St Helens (USA) in 1980 and Mt Pinatubo (Philippines) in 1991, were dramatic and the images captured at the time are certainly awesome and frightening. Mt St Helens ejected half a cubic kilometre of material into the atmosphere; on film, it looks like a lot but really, it is but a

---

[59] Best known for his "The Story of Civilization"; 11 volumes written in collaboration with his wife Ariel Durant.

volcanic pop gun. Mt Pinatubo was substantially larger, ejecting nine times as much volcanic debris, or 4.5km^3; a bigger pop gun but a pop gun still.

Indonesia has been home to many giant eruptions, whose impact on the wider world is impossible to ignore. The famous eruption of Krakatoa, between Sumatra and Java, in 1883, was heard in Australia and killed an estimated 36,000 people. It was a big eruption, spilling around 12km^3 of material into the atmosphere and dropping temperatures in the next northern summer by over one degree Celsius. More than a pop gun it is true.

Earlier in the 19th Century, in April 1815, a much larger eruption took place on the island of Sumbawa when the volcano Tambora erupted, ejecting 80km^3 of ash and dust into the atmosphere and causing global temperatures to fall significantly. The following year, 1816, was known as the "Year Without a Summer". There was widespread famine in Europe; food riots occurred in Germany; crops failed in North America; heavy snow fell in June in upstate New York; typhus, cholera and other diseases broke out in India and Asia; there were huge floods on the Yangtse River in China; tens of thousands of people died from either starvation or disease. But there was at least one saving grace – the dust and sulphur aerosols ejected into the high atmosphere by the eruption provided spectacular sunsets that were captured on canvas by the famous English artist, Joseph Turner. But even Tambora pales into insignificance when compared with Toba.

It has been estimated that 2,800km^3 of ash, rock debris and dust were blasted into the atmosphere by Toba in an eruption 5,600 times larger than the 1980 event at Mt St Helens. Its impact on climate and on all living things would have been utterly devastating, worldwide in reach and probably lasted for years, perhaps decades or more. Our forefathers, prehistoric *Homo Sapiens*, almost died out, reducing to perhaps fewer than 30,000 individuals[60].

Earthquakes are also to be feared and can cause major loss of life – witness the Boxing Day earthquake and tsunami just off Sumatra. But tragic and devastating as that event was, its effects were still essentially local. Not so the next giant eruption from a supervolcano. Whether in Indonesia, or perhaps America (Yellowstone), New Zealand (Lake Taupo), or Italy (Campi Flegrei) it will be much more devastating than anything experienced in recorded history. Its impact will be global, inescapable, uncontrollable. In a world where over seven billion people already crowd together and compete for food and water, civilisation will be severely threatened. Famine will be almost universal; air travel will cease; communications will break down; water supplies will be contaminated and crops will fail. That such an eruption will happen is certain. When it will happen is unknown.

So how might we respond to this rather bleak though inevitable and esoteric threat? By now my own view should be obvious: Expand your vision to truly see our world and not just look at it, for there is always more to the story of our planet than meets a superficial gaze. Appreciate the earth for what it is today; explore its idiosyncrasies; acknowledge its complexity; study its history; celebrate its wonders; and be aware that it is a dynamic, self-renewing system, vastly more changeable and unpredictable that we imagine. We humans are given but a fleeting glance of an evolving, living entity whose timeframe mocks our own sense of antiquity.

The book of Genesis in the Bible encourages us to "possess the earth and subdue it". The first bit we have done; the second bit we will never do. We exist despite, but also because of, that living entity and

---

[60] Not all researchers agree that the Toba eruption was the sole cause of this so-called "human bottleneck".

we should be grateful for that. Protect our world, cherish it but above all enjoy it, for it will not always be the same. The ancient Delphic maxim *"Know Thyself"* might reasonably be extended to *"Know Thy World"* for we are part of that world and we cannot truly understand our place in the cosmos without understanding the earth beneath our feet.

For me, geology has been both a career and a source of meaning for my existence. Comprehending something of this planet, the length of its history and the resilience of its processes has enriched my life. And paid the bills along the way! It has come to me as air, as fire, as water, but most of all as earth, from which I came and to which I will return. In one sense, my journey through the elements is nearly complete and yet, if I have learned anything over the past seven decades it is this: Nothing is permanent, no river, no lake, no mountain, not even a sea or a continent and certainly no person. But equally, nothing ever really vanishes completely. Continents form, break up and re-form. Mountains grow, are eroded and grow again. Life for the individual, even for a species, comes and goes, but there is always another to take its place. I am the stuff of stars, I know my place in the cosmos and most of all, I am my Country and that Country is me.

If sharing with you one man's journey through the elements has awakened your senses to this planet and our place upon it then my journey is truly complete. For I will leave with you a legacy of appreciation for our Earth that will in turn enrich your life and endow you with some of the awe, respect and meaning that have empowered my life experience and made me who I am.

Touring volcanic terrain in Iceland with renowned Icelandic vulcanologist, Haraldur Sigurdsson, in 2015

Inside the crater at White Island, New Zealand's most active volcano, in February, 2017

Karymsky volcano in Kamchatka in 2012

ACKNOWLEDGEMENTS

There are many people to whom I owe thanks for their roles in enabling and enriching the experiences related in this book.

This memoir has benefited greatly from the insightful critiquing, chapter by chapter, of my writing group friends, for which I thank them very sincerely. I also thank Prof Ian Plimer for his support and encouragement, including reading the draft manuscript, and Prof Ross Large for generously agreeing to write the Foreword.

I have worked with a lot of fine colleagues, who have inspired and encouraged me through all my endeavours and made my life both more interesting and more rewarding. Where appropriate, I have referred to them by name; in other places, I have used just first names or fictitious names because I can't remember exactly who it was that shared those experiences with me, or I thought anonymity was the wiser course. If that offends anyone I sincerely apologise and I thank them all, whether named or not.

Several key people stand out in the long list of *Dramatis Personnae* to whom I owe special thanks.

The first is the late Ian Carmichael, my indomitable Professor, who taught me that striving for the best is not an option, it is a fundamental attribute of a worthwhile life. His inspiration and the standards he set for me in the 1960s have continued to direct and guide me through life ever since.

The second is the late Lindley Sale, who showed me what a true gentleman looks like. He also, I strongly suspect, facilitated the grant from the Ford Foundation that made the PNG and PhD experiences possible.

My parents, Wally and Coral, now deceased, were always there for me, encouraging, supporting and working hard to allow me to pursue my dreams. It was my aviator older brother, Graham, who first spiked my fascination with volcanoes, with his stories from flying in Papua New Guinea in the 1950s and early 1960s. My younger brother, Brian, helped keep my feet on the ground when I got a bit carried away with self-importance and his skill as a golfer, continuing to this day, has been a constant inspiration and source of family pride.

But most of all, I thank my wife Margaret, without whose love and support none of these experiences would have happened. It was Margaret's diary, kept daily while we were in Papua New Guinea, that enabled me to provide such detail about names, dates and events as I have and I thank her for permission to use that diary as a major source for the second part of this manuscript. I also thank her for her perseverance and fortitude, as I travelled far and wide, often for extended periods, leaving her to keep the home fires burning and raise our three wonderful children, Julianne, David and Megan. To them I apologise for my frequent absences, for as much as a third of their formative years, but I hope I have been able to make it up to them in other ways.

Perhaps this book will show them that their forbearance was worthwhile.

www.ingramcontent.com/pod-product-compliance
Lightning Source LLC
Chambersburg PA
CBHW052056300426
44117CB00013B/2146